Robert Lacoste's the Darker Side

Practical Applications for Electronic Design Concepts

Robert Lacoste

ELSEVIER

AMSTERDAM • BOSTON • HEIDELBERG • LONDON
NEW YORK • OXFORD • PARIS • SAN DIEGO
SAN FRANCISCO • SINGAPORE • SYDNEY • TOKYO

Newnes is an imprint of Elsevier

Newnes

Newnes is an imprint of Elsevier
30 Corporate Drive, Suite 400
Burlington, MA 01803, USA

The Boulevard, Langford Lane
Kidlington, Oxford, OX5 1GB, UK

Notices

Knowledge and best practice in this field are constantly changing. As new research and experience broaden our understanding, changes in research methods, professional practices, or medical treatment may become necessary.

Practitioners and researchers must always rely on their own experience and knowledge in evaluating and using any information, methods, compounds, or experiments described herein. In using such information or methods they should be mindful of their own safety and the safety of others, including parties for whom they have a professional responsibility.

To the fullest extent of the law, neither the Publisher nor the authors, contributors, or editors, assume any liability for any injury and/or damage to persons or property as a matter of products liability, negligence or otherwise, or from any use or operation of any methods, products, instructions, or ideas contained in the material herein.

Library of Congress Cataloging-in-Publication Data
Application submitted.

ISBN: 978-1-85617-762-7

British Library Cataloguing-in-Publication Data
A catalogue record for this book is available from the British Library.

For information on all Newnes publications
visit our Web site at *www.elsevierdirect.com*

Working together to grow
libraries in developing countries

www.elsevier.com | www.bookaid.org | www.sabre.org

ELSEVIER   BOOK AID International   Sabre Foundation

Contents

Foreword

While I would like to tell a story about how *Circuit Cellar* magazine made Robert Lacoste into the respected authority on electronic engineering that he is today, I do not pride myself a thief. His success has been his own doing. I have never seen a man so skilled at his craft, yet so willing to work diligently to demystify what is often referred to as black magic electronics engineering. Circuit Cellar owes Robert its sincere gratitude for allowing its pages to be his venue of choice over the last several years.

Having acknowledged this, I must to note that this book represents one of the finest examples of Circuit Cellar's evolutionary success. I first became aware of Robert's work through his project submissions to Circuit Cellar's embedded design contests. His résumé of contest success was so impressive, covering a wide range of expertise and knowledge of microcontrollers, that after nine contest entries and seven top-four placements (including one Grand Prize and multiple First Prizes), Robert became a third-party judge for future contests. The work of a design contest judge is grueling. Each project must be studied in tremendous detail and evaluated against a host of abstract and concrete design criteria. It's a job someone who cherishes the spirit of innovation in the design community. I know that Robert wears the title "*Circuit Cellar Contest Judge*" as badge of honor.

Robert quickly became a recurring feature contributor to *Circuit Cellar*'s magazine. Even his earliest contributions from over a decade ago reveal a drive for going above and beyond the requirements of the job. Robert was the teacher who cared about the student. His articles were the tools that would inspire everyone, from seasoned pros to design students.

Robert's willingness to tackle the most difficult-to-understand topics in electronics engineering and his ability to demystify them through his writing and project examples, made it very easy for him to go even further in his Circuit Cellar evolution. He was welcomed aboard as a bi-monthly columnist in August 2007 and has since presented much-acclaimed pieces under the heading "The Darker Side." The feedback from readers has been extremely positive. It's a joy to hear from those with decades of experience in embedded development and discover how much they're learning. In "The Darker Side" column, *Circuit Cellar* readers have a resource for thought-provoking design concepts. In *Robert Lacoste's the Darker Side,* the book you now hold in your hands, the embedded design community has a collection that is truly worthy of any developer's library.

Thanks, Robert, for being a part of the *Circuit Cellar* evolution success story.

Steve Ciarcia
Founder/Editorial Director
Circuit Cellar *Magazine*

Preface

An engineer always has the same difficult choice when placed in front of a new project. Is it better to rely on his knowledge and experience and to use solutions that he has already used somewhere else, or should he extend his knowledge and take a more risky route, meaning to learn and to try new and innovative techniques?

I call the first approach the "comfort zone" solution. It is of course a very reasonable one, and I hope it is the preferred method for most safety-critical systems designers. It is also the preferred method of project managers who know that the schedule will be less at risk. Staying in your comfort zone means that you probably will have no bad surprises. However, the downside is … that you will never have good surprises. Your product will not become significantly better than your competitor's, its price will not go down drastically, and your sales will not double.

On the contrary, the innovative approach has downsides too, and nobody should deny or ignore them. You may find more difficulties than anticipated, or even the solution you have tried may appear purely inadequate for the specific project. More frequently, your schedule may be in jeopardy and you may have some difficulties with your management. However, if you are clever (or lucky), then impressive achievements are possible. You could end up with a really disruptive solution, providing either better performances, more flexibility, more features, or lower cost than existing solutions. This means that you will please your customers or boss. And even if you are not designing something that will change our lives completely, this also means that you will leave your small contribution to humanity.

Personally I think life is too short to stop learning. Of course risks must be mitigated, but I always ended up with a frustrating feeling when I was only staying in my comfort zone. Honestly I spent the first fifteen years of my professional life in a quite schizophrenic mode. During working hours I managed huge technological projects in information technologies and telecommunications, meaning of course that I became farer and farer to technology and more and more reluctant to risks. To compensate I spent my nights developing innovative electronic devices as a hobbyist, using new techniques each time and especially focusing on things that I had never learned or tried. The design contests organized by *Circuit Cellar* were very helpful motivators to develop as impressive as possible solutions periodically!

Back in 2003 I decided that this life was not as optimal as possible, so I quit my last employer to launch my own consulting and design small company, ALCIOM. My goal was simply to bring the happiest part of my activities, meaning innovative design work, back into my daily time. It was a very fortunate decision. We are now working 70 percent for start-up companies who clearly ask us to use as innovative solutions as possible and 30 percent for large companies who are asking us to help them to go back to a start-up like approach. With my colleagues we are doing our best to keep the same spirit over the years, meaning we try to use new and innovative solutions in every project, even if it is only a small subpart of a project. This definitely increases our own comfort zone over time, makes us more proud of our lives, and we hope makes customers happier.

When Steve Ciarcia asked me to write a new bi-monthly column for *Circuit Cellar*, I immediately thought that the best idea would be to help readers go out of their own comfort zones. This means to present some of the lesser-known, more obscure aspects of electronic design, to highlight concepts too often perceived as expert-only subjects (if you read "analog" between these words, you will probably be right 90 percent of the time). I also wanted to explain what was really going on and to focus on application-oriented explanations and pragmatic tips rather than painful theory and math (although some equations may show up from time to time). The Darker Side column was born. As I got very positive feedback from readers, I continued with articles covering subjects from electromagnetic interferences to antennas, from digital filters to phase-locked loops, or from digital modulations to control systems.

This book is a collection of Darker Side columns. Because space constraints are easier in a book than in a magazine, I added plenty of additional examples and explanations, as well as a couple of exclusive chapters. Each chapter is nearly fully independent from the others, so don't hesitate to jump directly to a specific section if you are in a hurry and have a short-term problem with the project on your desk. Two chapters deal with notions used in other sections, namely Chapter 1 on impedance matching and Chapter 6 on Fourier transform, so you'd better start with these two if you are not already fluent with these notions. You also will find some additional informations in the appendixes, in particular, appendix A gives you a tutorial on a tool that I used extensively through out this book, namely SciLab.

Anyway even if the chapters are independent of each other, I still suggest that you take your time and read the book from start to finish, maybe jumping over chapters where you are already in your comfort zone. This would help you discover that some techniques could be useful for your next project, either if you don't already know what it is. Plus, this will help you reduce your own darker side. That's my target!

Finally, don't hesitate to contact me by email. The topic I'm presenting in this book was indeed out of my comfort zone no long ago, so I may be wrong or imprecise from time to time. Moreover you may also be able to reduce my remaining darker side!

Acknowledgments

I would like to dedicate this book to all the people who helped me do what I love to do. They are really numerous, but here is a short list. To my father, who taught me how to solder my first transistor. To my mother, who supported my father and myself when we brought trashed TVs back home to dismount them in the living room. To Steve Ciarcia who didn't know at the time that I was reading every word of his articles from *Byte* magazine to *Circuit Cellar*, and especially as my first microcontroller-based project was built following one of his papers. To all the great *Circuit Cellar* team, who continuously encouraged me and helped me step by step to correct my Frenchy English. To the *Circuit Cellar* contest sponsors, who pushed me, as well as thousands of other engineers, to find new ideas and to make them work. To Elsevier, who pushed me to transform these materials into a book. To my colleagues at Alciom, who took the risk to join me in my consulting adventure. To our customers, who trusted us and gave us fantastic innovative projects to work on. And, of course, to my dear wife Isabelle and to our daughters Pauline and Adele, who patiently forgive my long nights spent with a soldering iron in hand or in front of my PC writing this book, rather than repairing the garage door.

Part 1
Impedance Concerns

CHAPTER 1

Impedance Matching Basics

Let's start our exploration of the darker side with a quite simple but often misunderstood topic: impedance matching. As we will see, this concept is fundamental for all radio and small signal designs, basically because radio signal power is expensive and must be well used. Just try to replace your 50-m 75-ohm TV antenna cable with a 50-ohm cable to understand what I mean: You may be disappointed with the image quality. However, matching is now also fundamental in nearly all high-speed digital designs. Our standard PC motherboard, with its GHz clock, will simply not work at all if impedance matching was not managed properly by its designers. In that case, the problem would not be one of efficient energy use but one of signal reflection linked to impedance mismatches. We will see an actual demonstration of this phenomenon in Chapter 3 on time domain reflectometry. Antennas are also fundamentally related to matching. Basically, an antenna is roughly a matching network between an electrical signal and space, but more on that in Chapter 14.

Optimal Heating for the Winter

Okay, let's take the most basic example first. Suppose that you are in the middle of a deep forest during a cold winter's night. You still have some coffee, but it is, unfortunately, really cold. Because you are an electrical engineer, you always have a 12-V battery pack in your bag, as well as plenty of power resistors and extra wires. Bingo, you could easily build a heater for your coffee just by connecting a resistor R

Note: This chapter is partially based on the column "The darker side: Antenna basics," *Circuit Cellar*, no. 211, February 2008.

© Elsevier Inc.
DOI: 10.1016/C2009-0-20196-6

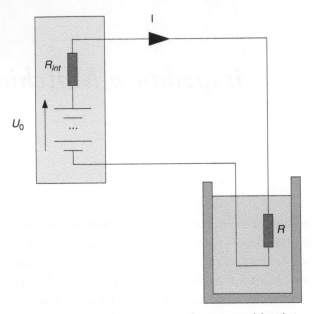

Figure 1.1: The most basic impedance matching issue.

across the terminals of the battery and putting this resistor in your coffee. The resistor will heat up, and so will your coffee. Assuming there aren't any thermal losses, the power dissipated by the resistor will be fully transformed into a coffee temperature increase. But what would be the optimal value for this resistor R in order to use your battery's energy in the most efficient way—that is, to get as much energy as possible from it? This is exactly an impedance-matching problem (Figure 1.1)

Since no perfect voltage sources exist, any so-called fixed voltage supply has at least an internal parasitic serial resistance R_{int}. Suppose that your battery pack has an internal resistance of 1 ohm. This means that the output voltage of the battery will be $U_0 = 12$ V with no load; 11 V with a 1A load, as the voltage drop on the 1-ohm internal resistance will be 1 V when the current is 1A; 10 V with 2A; and so on. And of course let's assume that you can't change the internal resistor of the battery. The application of Ohm's law is then straightforward.

The battery is loaded with two resistors in series, R_{int} and R, so the current going through the circuit loop is

$$I = \frac{U_0}{R_{total}} = \frac{U_0}{R_{int} + R}$$

The power dissipated in your coffee will then be

$$P = R \cdot I^2 = R \cdot \left(\frac{U_0}{R_{int} + R} \right)^2 = \frac{R}{\left(R_{int} + R \right)^2} \cdot U_0^2$$

What are your choices? If you use a high value for the resistor R, then the current going through it will be small, and therefore the dissipated power will be low—see the preceding equation. If R increases, then the denominator will increase faster than the numerator and the power will go down to zero if R is very high. Conversely, if you use a very-low-value resistor (close to a short circuit), then the current through the circuit will be very high; however, the power dissipated in your resistor will also be very low because the voltage across a short circuit will be nearly zero. The preceding equation also shows that P will be zero if R is null. In fact, in this last case, a lot of power will be dissipated, but only in the battery's internal resistance and not in your heating resistor. So there must be an intermediate optimal resistor value for maximum power transfer from the battery to your resistor.

Let's plot the power P, defined by the previous equation, when the value of the loading resistor R varies. This is also a simple but good opportunity to present one of the simulation tools I will use throughout this book, Scilab, which is a French numerical calculation tool similar to software such as MathLab (© The Mathworks) but available freely on an open-source license basis. You will find a tutorial on Scilab in Appendix A, but its syntax is quite straightforward for such a simple plotting example:

```
Rint=1;                 // internal resistor = 1 ohm
U0=12;                  // Battery open voltage = 12V
R=[0:0.05:10];          // R range from 0 ohm to 10 ohm with
                           0.05 ohm steps
P=R./(Rint+R)^2*U0^2;   // Calculate the corresponding power
plot2d(R,P);            // Plot it
```

If you run Scilab, you will find that there is actually an optimal value for R. This value is 1 ohm, which is the same as the source resistance. For those who remember how to calculate a derivative, you can also derive the equation given P, and you will find that

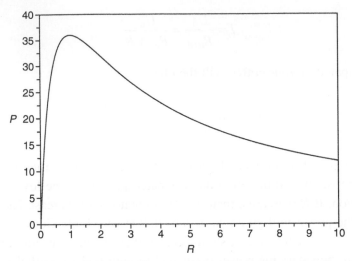

Figure 1.2: The power P dissipated in the resistor is maximum for $R = 1$ ohm = R_{int}.

this derivative is null for $R = R_{int}$. The maximum power dissipated in R is then 36 W, and exactly the same power of 36 W is dissipated in the battery's internal resistor R_{int} (Figure 1.2).

So we have found that the optimal value of the loading resistor, meaning the value that allows as much power as possible to be obtained from the source, is *the same resistance as the internal resistance of the source*. This result is the fundamental theorem of DC impedance matching. Keep in mind that it is applicable any time that a power transfer must be optimized. That doesn't mean that you must always load a battery with a matching load, but you should if your goal is to draw as much power as possible from the source to the outside world. Be careful—a common misinterpretation of impedance matching is trying to see it as a way to "save" energy or power. It isn't. Impedance matching is a way to get as much power as possible from a source that has a given, not-null, internal resistance.

What About AC?

If we move from DC to AC, the situation is exactly the same: replacing resistance with impedance. If you have a source with a given impedance Z_S and connect it to a load with a given impedance Z_L, then the power transfer will be maximum if both impedances are matched. There are just two differences in AC. An impedance Z is a complex value—the sum of a real resistance R and an imaginary reactance X. These impedances can be written respectively:

$$Z_S = R_S + jX_S$$

and

$$Z_L = R_L + jX_L$$

Do not be intimidated by this complex notation; just consider it an easy way of managing a pair of different values, R and X, as a single "complex" number written as $R + jX$. If you wish, refer to the short introduction on complex numbers provided in Appendix B.

With these notations, it can then be easily shown that matching is achieved when the load impedance is the *complex conjugate* of the source impedance. This simply means that both resistances should be equal, as in the DC case; however, both reactances should be equal in value but with opposite signs. Recall that a reactance is positive if the circuit is inductive at the working frequency, and negative if it is capacitive. In other words, a slightly capacitive source must be matched with a slightly inductive load and vice versa. The AC matching condition can then be written as

$$Z_L = \overline{Z}_S \Rightarrow R_L = R_S \text{ and } X_L = -X_s$$

The other difference between DC and AC is that the reactance of a load changes with the frequency. For simple capacitors and inductors the well-known formulae are the following:

$$X_C = -\frac{1}{2\pi f C}$$

$$X_L = +2\pi f L$$

where C is the capacitance in farads, L is the inductance in henries, f is the frequency in hertz, and X is the reactance in ohms.

The consequence of this reactance frequency dependency is that matching is also usually frequency dependent: A given load impedance can provide a good matching at a certain signal frequency, but this will usually not be the case at another frequency. That makes life funnier.

Matching Issues

Let's look at an AC mismatch example. Figure 1.3 shows a simulation of a serial RLC network done with the Quite Universal Circuit Simulator (QUCS), an open-source broad target simulator software project headed by Michael Margraf and Stefan Jahn. Such an RLC network has a zero reactance at its resonant frequency, so, at that precise frequency, its impedance is equal to the serial resistance, R, which is 20 ohms in this example. We expect to have a maximum power transfer at this resonant frequency, but not a perfect power transfer because this resistance is different from the 50-ohm source resistance. In this simple example, the resonance frequency can be calculated as the inverse of $\sqrt{L \times C}/2\pi$, which equals 1.59 GHz. This is exactly what is shown in the simulation.

Suppose that you have such a 1.59-GHz RLC resonant circuit that you can't change but that you need to connect to a Wi-Fi transmitter (2.4 GHz), which has a 50-ohm output impedance like our generator. Because you are far from the resonant frequency, if you connect them directly, you will have mismatch losses. More exactly, Figure 1.3 shows that you will get only around 0.35 W or 35% of the available power.

LC Matching Networks

What can you do? The answer is shown in Figure 1.4: You can add a matching network.

In this example, the matching network is made with another coil and capacitor, forming a low-pass filter, and, voila, you get a 100% power transfer at 2.4 GHz without changing the load. Basically, the *LC* matching stage just takes care to show a different complex impedance on both sides without any losses except through the internal parasitic resistances of the components. As shown in Figure 1.4, the power transfer is now optimal at 2.4 GHz, but only at that frequency; such an *LC* matching network is narrowband.

Of course, the values of the *LC* network I used in this example were calculated to get a matching condition at 2.4 GHz, that is, to ensure that the overall impedance of the load,

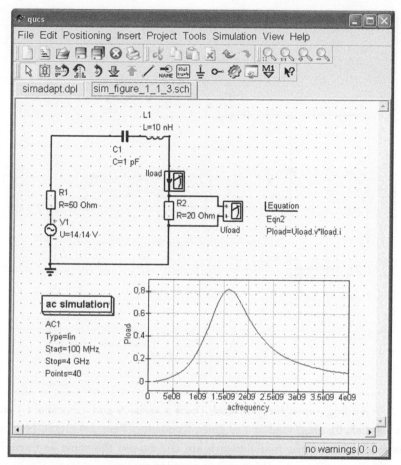

Figure 1.3: This QUCS simulation shows an RLC resonant network. The Uload and Iload meters measure the voltage and current across the load, respectively. The equation calculates the product, which is the power dissipated in the load. The graph shows that the dissipated power is maximum (a little more than 0.8 W) at the resonant frequency. The power transfer is not optimal because a matched 50-ohm load would allow 1 W to be obtained from a 14.14-Vrms generator, which has an internal resistance of 50 ohms. When the maximum is achieved, the same power (1 W) would also be dissipated inside the generator.

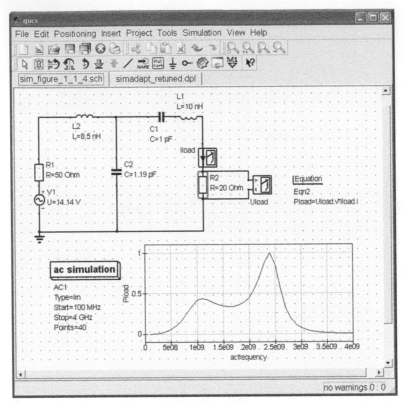

Figure 1.4: Adding a matching network (here a serial *L* and a parallel *C*) enables you to match the *RLC* load at any given frequency (here 2.4 GHz). Note that the *RLC* values of the load network are still exactly the same as those in the previous simulation.

plus the matching network, is equal to the complex conjugate of the impedance of the source. But how can you calculate these components? Because you already know the impedance of the source (50 ohms), the first step is to know the impedance of the load at the desired 2.4-GHz frequency. You can calculate it, measure it, or simulate it. Using QUCS, just add a voltmeter and an amp meter to measure the current and voltage delivered by the source, divide them ($Z = U/I$), and simulate the network at 2.4 GHz. You get the impedance, which is $20 + j \times 84.5$ ohms (Figure 1.5).

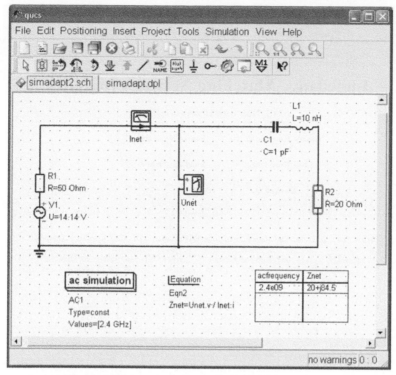

Figure 1.5: QUCS allows you to calculate the load impedance at 2.4 GHz. Just measure voltage and current, and calculate their ratio.

The reactance is positive, +84.5 ohms, so the circuit behaves like an inductor at this frequency, which is usually the case for frequencies above the resonant frequency. You then have to calculate a network that can match 50 ohms to $20 + j \times 84.5$ ohms at 2.4 GHz. You can use the good old methods that you will find in your nearest library or use hand calculation or an abacus. Or, if you are like me, you can simply use a matching calculator. Let's use the online "Matcher2.htm" tool from John Wetherell that you will find on the Web site of the University of San Diego (*http://home.sandiego. edu/~ekim/e194rfs01/jwmatcher/matcher2.html*) (Figure 1.6).

As you can see, these component values were the ones used in the Figure 1.4 matching network.

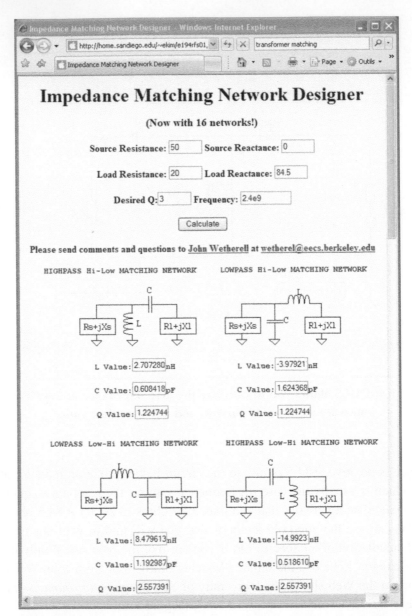

Figure 1.6: Calculation of a simple *LC* matching network. Several topologies are possible and will give the same matching. I used the bottom-left solution for Figure 1.4.

Matching and Reflections

Another way to understand an imperfect matching condition that is especially useful at high frequencies is to consider that some part of the power sent by the generator to the load is "reflected back" to the source if the matching condition is not achieved. The amplitude and phase of the reflected signal add to or subtract from the incident wave, canceling part of the energy transfer. This is actually the case even if it seems strange, as we will see in Chapter 3 on time domain reflectometry. Just as an introduction: Imagine that you switch on the source. Because electricity moves only at the speed of light, it cannot instantly "know" if the load is matched or not. Power needs to go through the cable to the load to discover that the load is not matched and then send back a reflected wave to the source to say, "Hey, it's not properly matched." This is a very simplified explanation, but it is enough to prove that matching is important. These reflected waves can have other negative impacts on RF designs (standing waves on the cables, nonlinear distortion in the amplifiers, and so on), and we will devote Chapter 3 to this topic.

I needed to talk about reflection to introduce an important concept: S parameters. In its simplest form, the voltage ratio between the incident and the reflected wave is noted as S11. The amplitude of this coefficient is 0 if the network is matched (no reflection). It is 1 if it is completely unmatched (100% reflection). This is also a complex value (amplitude and phase of the reflected signal), and it can be plotted either on an *XY* scale or on a polar form, which is named the Smith graph. Of course, QUCS can calculate S11 coefficients (Figure 1.7).

The Smith diagram is an invaluable tool, as it allows matching networks to be calculated graphically, and it was the only solution before computers. However, this topic would need another chapter.

Any Other Solution?

Such simple *LC* matching networks are very efficient, but they have a fundamental limitation: They are narrowband. You will not be able to use such a network if you need to match a source to a load over a large frequency range. Furthermore, you will quickly discover that the matching is narrower when the difference between load and

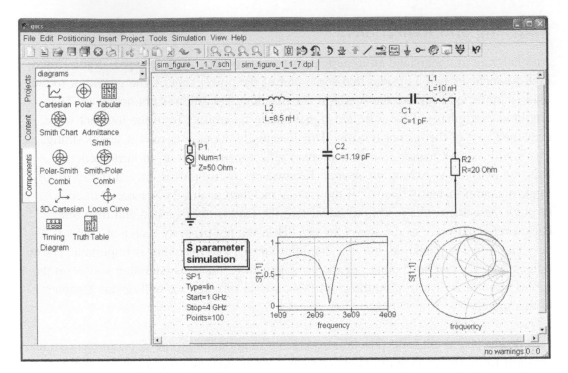

Figure 1.7: These plots show the reflection coefficient (S11) of the matched network illustrated in Figure 1.3 in a polar form (Smith diagram) and on an *XY* plot. Matching is achieved when S11 is null (at the center of the plot in a polar form). A Smith chart shows when the network is inductive (top half) or capacitive (bottom half), and it can be used to calculate matching networks manually.

source impedances is larger. You will also discover that the calculation of the matching network can give unreasonable values such as 0.01 pF or hundreds of henries if the source and load impedances are very far away, even if there is always a theoretical matching solution.

What are the alternatives? The first type of solution is to use more complex *LC* matching networks, which can allow either the use of more reasonable component values with other network topologies or the expansion of matching bandpass.

For wideband requirements other ideas can be used. Since such a network is fundamentally a filter, you can use several matching filters in parallel, one per frequency band. This works, but it may be quite difficult to optimize as some cross-coupling between filters will inevitably happen unless they can be switched. However, don't forget the other ideal wideband matching network: the old but trusty signal transformer. Effectively, a transformer with a turn ratio of $1:N$ (N times more turns on the secondary than on the primary) is an impedance transformer from any impedance R to an impedance $N^2 \times R$. It is a very wideband solution, even for three frequency decades or more; however, it does not help if the load is not a pure resistance: If the load has a non-null reactance, then you will first need to cancel this reactance with a frequency-dependent matching network. Anyway, transformers can be a tremendous help for difficult matching situations.

Finally, resistors can be used as impedance matchers. However, resistor-based matching stages are usually not used just because they are intrinsically lossy, and a matching network is there to use power efficiently. Anyway, a resistive attenuator may be a not-so-bad solution in case of excessive reflections when no other techniques are possible, in particular for digital designs.

Wrapping Up

Now you know why impedance matching can make the difference between a working and a nonworking project, especially in RF designs where signal power is an expensive resource. In a nutshell, the power transfer between a source and a receiver is maximized when impedances are matched. Remember, this means that the source impedance is the complex conjugate of the load impedance: Both resistances must be equal and both reactances must be equal in value, but with opposite signs. If the source impedance is a 50-ohm pure resistance, matching will be achieved when the load is also a pure 50-ohm resistance. And if a cable or wire is used between them, it must also have a 50-ohm characteristic impedance.

If the matching is not perfect, you won't get as much power as you could: Some power will be dissipated somewhere else than in your load. It is usually reflected back to the source and dissipated, but it could also generate negative effects such as distortion or spurious signals.

If you are working on RF designs, then 90% of the time you will try to keep a 50-ohm characteristic impedance through your circuit, or maybe 75 ohms if you are working on video. By the way, do you know where these usual 50- and 75-ohm values come from? I discovered the explanation in Thomas H. Lee's book *Planar Microwave Engineering: A Practical Guide to Theory, Measurement, and Circuits*. According to Lee, for a given coaxial cable diameter, there is a precise ratio of inner to outer conductor size, which gives the minimum intrinsic resistive loss. This ratio corresponds to a characteristic impedance of 77 ohms. That's why a close value, 75 ohms, is used for video where signals are small and where attenuation should be kept as small as possible, even on long cables.

Okay, but why 50 ohms elsewhere? Because there is another optimization to deal with. A coaxial cable can transmit a given maximum peak power, corresponding to the dielectric breakdown of the cable due to high RF voltages. It happens that, for a fixed external size, this power-handling capacity is maximized with another ratio of inner to outer conductor size, corresponding to a characteristic impedance of 30 ohms, at least when air is used as a dielectric. Back in the 1930s, engineers hesitated between 77 and 30 ohms and chose an intermediate value for common coaxial cables, 50 ohms. To be honest, there are other explanations on the Internet, but this one seems plausible.

We can devote entire books to impedance matching, but unfortunately I have other darker side subjects to share with you. I hope that these explanations gave you some basic ideas on matching, and I hope that they were also a good refresher for expert readers who will forgive me for the simplifications used in this chapter. Anyway, let's now move to other aspects that are still linked to impedance matching.

CHAPTER 2

Microstrip Techniques

With this chapter, the title of this book has never been more appropriate because I will talk about something that may really seem magical to the novice. Follow along as I introduce microstrip techniques and explain how to implement zero-cost components by simply drawing PCB tracks.

Transmission Lines

Before digging into microstrips, we need to spend a minute on transmission lines. In Chapter 1, I explained impedance matching and briefly said that the connection between a 50-ohm source and a 50-ohm load must be made using a cable with a 50-ohm characteristic impedance for proper matching. The impedance of a load or source is quite straightforward to understand, but what actually is the characteristic impedance of a cable?

Imagine that you have a given length of coaxial cable and you divide it into a large number of very small sections, each of length dZ. If you ignore parasitic resistances, then each section can be modeled as a small serial inductance, because any wire has a non-null inductance, and as a small capacitor between the wire and the ground, because the central wire is not far from the grounded shield. These parasitic inductances and

Note: This chapter is a corrected reprint of the article "The darker side: Microchip techniques," *Circuit Cellar*, no. 223, February 2009.

© Elsevier Inc.
DOI: 10.1016/C2009-0-20196-6

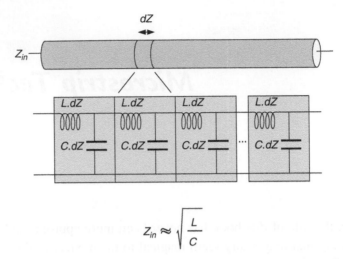

$$Z_{in} \approx \sqrt{\frac{L}{C}}$$

Figure 2.1: A transmission line can be modeled as a succession of small L/C networks. The mathematical relationship between the L and C parameters and the characteristic impedance of the line is simple to apply and slightly more complex to demonstrate.

capacitances are roughly proportional to the length of the small section, so they can be noted $L.dZ$ and $C.dZ$, with L and C in henries per meter and farads per meter, respectively. Therefore, the wire can be approximated as a serially connected set of identical $L.dZ/C.dZ$ networks (Figure 2.1).

If you apply a DC voltage step U on one end of the cable, some current will flow through the cable until all small capacitors are charged. If you have an infinite length of cable, this current will flow forever and will quickly stabilize to a given value I. The ratio U/I is homogenous to a resistance. Similarly, if you apply an AC voltage to the input, you will get an impedance. This impedance is what is called the cable's characteristic impedance: The characteristic impedance of a cable is the impedance it would have if it were infinite. In fact, it can be quite easily demonstrated that this characteristic impedance is simply the square root of L/C (see sidebar).

For those interested in the demonstration, just consider that you "cut" an infinite cable of impedance Z_0 just after the first LC section. The rightmost section is still an infinite length of cable, so it should also have an impedance of Z_0. So Z_0 must be equal to the impedance of L in series with the assembly of C and Z_0 connected in parallel. Let's write this:

$$Z_0 = Z_L + \frac{Z_0 \cdot Z_C}{Z_0 + Z_C} = j\omega L + \frac{Z_0 \cdot \dfrac{1}{j\omega C}}{Z_0 + \dfrac{1}{j\omega C}}$$

$$= \frac{Z_0 + j\omega L - \omega^2 L C Z_0}{j\omega C Z_0 + 1} \approx \frac{Z_0 + j\omega L}{j\omega C Z_0 + 1}$$

Solving this equation will quickly give you the solution:

$$Z_0 (j\omega C Z_0 + 1) = Z_0 + j\omega L$$

so

$$Z_0 \, j\omega C Z_0 = j\omega L$$

and then

$$Z_0 = \sqrt{\frac{L}{C}}$$

Anyway, this demonstration was just there for your pleasure.

Just remember for the moment that a cable, like any transmission line, can be modeled as a succession of small LC sections and that the characteristic impedance of the cable increases with L and decreases with C. Of course, you can't measure the characteristic impedance of a cable with an ohmmeter because your cable will never be infinite; nonetheless, it is an intrinsic characteristic of the cable, defined by its physical construction. For example, for a coaxial cable, this impedance can be calculated as shown in the following equation.

$$Z_0 = \frac{138}{\sqrt{\varepsilon_r}} \cdot \log \frac{R}{r}$$

where ε_r is the relative permittivity of the cable's dielectric, R is the inner radius of the outer conductor, and r is the radius of the inner conductor.

Microstrip Tracks

Now what happens if you do not have a coaxial cable but simply a copper track on one side of a PCB with a complete ground plane on the other side? The situation is the same as it is with the coaxial example. The track can be split into small sections, and each section can be modeled the same way. Thus, this track will have a characteristic impedance. Such a PCB track on a full ground plane is called a microstrip, which is the most common way to connect RF components on a PCB (Figure 2.2).

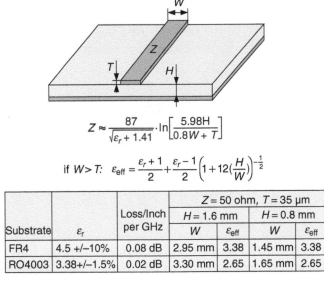

$$Z \approx \frac{87}{\sqrt{\varepsilon_r + 1.41}} \cdot \ln \left[\frac{5.98H}{0.8W + T} \right]$$

if $W > T$: $\varepsilon_{\text{eff}} = \frac{\varepsilon_r + 1}{2} + \frac{\varepsilon_r - 1}{2} \left(1 + 12 \left(\frac{H}{W} \right) \right)^{-\frac{1}{2}}$

Substrate	ε_r	Loss/Inch per GHz	$Z = 50$ ohm, $T = 35$ µm			
			$H = 1.6$ mm		$H = 0.8$ mm	
			W	ε_{eff}	W	ε_{eff}
FR4	4.5 +/−10%	0.08 dB	2.95 mm	3.38	1.45 mm	3.38
RO4003	3.38+/−1.5%	0.02 dB	3.30 mm	2.65	1.65 mm	2.65

Figure 2.2: A microstrip is simply a copper track running on one side of the PCB while the other side is a plain ground plane. The formula will give you the characteristic impedance of the track, as well as the effective dielectric constant based on the geometric parameters. The table provides usual values for 1.6- and 0.8-mm-thick PCBs, as well as values for the standard FR4 substrate or the most advanced Rogers RO4003.

You can use other settings such as a stripline, which is a track sandwiched between two ground planes on a multilayer PCB, but microstrip is the most frequently used option because it is affordable and well suited to SMT components.

How do you control the characteristic impedance of a microstrip track? The answer is with the track width, which is the only parameter that you can easily manage on your PCB CAD tool. Intuitively, if the track is wider, the capacitance between the track and the ground plane will increase and the characteristic impedance will decrease. If the track is thinner, its inductance and its characteristic impedance will increase. So there should be a given track width that corresponds exactly to 50 ohms, at least for a given PCB technology. This width is dependent on the PCB substrate (FR4 is the most common) through its dielectric constant and the PCB thickness (1.6 mm or 0.8 mm for double-sided designs), and it is slightly dependent on the copper thickness, which is 35 μm most of the time. The formula and the usual values are given in Figure 2.2, but there are good free calculation tools on the Internet, such as Agilent Technologies' AppCAD (Figure 2.3).

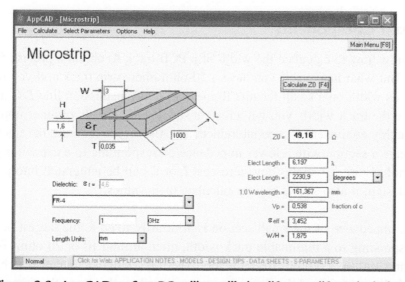

Figure 2.3: AppCAD, a free PC utility, will simplify your life. It includes a powerful microstrip calculator.

A 50-ohm track roughly corresponds to a 3-mm track on a standard 1.6-mm PCB, and to a 1.5-mm track on a 0.8-mm PCB. That's why it is often more appropriate to use 0.8-mm PCBs for RF projects, simply because the tracks have a more manageable width.

Let's summarize: Whenever you design a PCB for a high-frequency project, *you must always use tracks with the width for proper impedance matching and with a full ground plane on the opposite layer.* The only exception is when the length of the track is short compared with the signal's wavelength (e.g., a couple of millimeters long). In that case impedance matching may be neglected. A word of caution: Be careful if you use a multilayer PCB. The track width will need to be calculated based on the distance between the microstrip track and the ground plane, which is usually on the first inner layer. Ask your PCB supplier for the actual distance because this could be process-dependent. Also remember that the standard FR4 PCB substrate has a fuzzy dielectric constant (specified as ±10%) and high losses, which make it difficult to use when the working frequency exceeds a couple of gigahertz. Specific high-frequency substrates (e.g., Rogers RO4003) are much more efficient and well characterized, but they have a cost (see Figure 2.2).

Distributed Components

Now you know how to calculate the width of a PCB track to obtain a precise 50-ohm impedance. But what happens if you have a 50-ohm microstrip track and you increase or decrease its width on a small length? Remember the transmission line *L/C* model? If you decrease the track width, you will create a small section with a higher impedance, which is roughly equivalent to a serial inductance. And if you increase the track width, you will create a section with a lower impedance corresponding to a capacitor parallel to the ground! Figure 2.4 shows how zero-cost *L* or *C* can be integrated into a microstrip design; it also shows how to calculate their values.

Usually, the impedance of the small section is arbitrarily fixed to the largest or smallest value corresponding to a reasonable track width, often around 10 or 20 ohms for capacitors and 100 to 200 ohms for inductors. The track length calculation is based either on this hypothesis and the desired component value, thanks to the formulae supplied in Figure 2.4, or on the calculation tools.

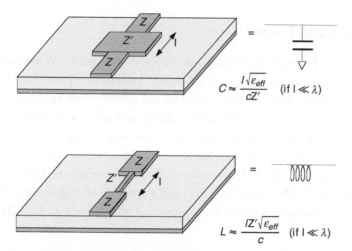

$$C \approx \frac{l\sqrt{\varepsilon_{eff}}}{cZ'} \quad (\text{if } l \ll \lambda)$$

$$L \approx \frac{lZ'\sqrt{\varepsilon_{eff}}}{c} \quad (\text{if } l \ll \lambda)$$

Figure 2.4: Parallel capacitors or serial inductors can be easily implemented on a microstrip line just by changing the track width, that is, the track impedance, on a precise length. Any reasonable impedance Z' can be used.

You need to be careful for two reasons. First, the dielectric coefficient used in the formulae is not the raw dielectric constant of the substrate found in the supplier's datasheets, but the effective dielectric constant of the microstrip wire, which is between the substrate and the air and exhibits a slightly different behavior. The formula and usual values are also provided in Figure 2.2. For example, the dielectric constant of FR4 is around 4.5, but the effective dielectric constant of a microstrip on FR4 is 3.38 because the track is not buried inside the FR4 material but lies at the border between FR4 and the air.

Second, you must remember that I took the hypothesis of "small sections" of a track for this discussion. What does it mean in practical terms? Simply, that *the L/C model will be erroneous as soon as the track dimensions (width or length) are not significantly small compared with the wavelength of the signal you are working on.* This wavelength is the speed of the signal on the PCB track divided by the working frequency. The signal speed is c (the speed of light, 3.108 m/s) divided by the square root of the effective dielectric constant. For example, this corresponds to a wavelength of 6.7 cm at 2.4 GHz on FR4. At that frequency, you can expect to have issues as soon as the component's dimension is larger than 1 cm or so. This fixes a limit to the value of the components you can design in microstrip form.

Need a Zero-Cost Filter?

Enough with theory. Now it's time for some funny experiments. What can you do with *L* and *C* devices? You can try to design a 50-ohm 1-GHz low-pass filter with absolutely no discrete components. Such a filter would be made only with specific copper tracks on the PCB, so its cost would be virtually nothing, at least if you consider PCB surfaces as free. The first step is to design the filter as if you were using classical lumped components. Because I'm lazy from time to time, I used a free online filter calculator developed by Tony Fisher at the University of York. The result in Figure 2.5 includes three inductances (9.12, 15.7, and 9.12 nH) and two 4.36-pF capacitors. The 3-dB cutoff frequency is precisely 1 GHz. The calculated attenuation at 1.5 GHz is around 20 dB.

You can build this filter using standard components. It will work if you are lucky enough to find a 9.12-nH inductance somewhere. How can you transform these values into microstrip components? Let's assume that the PCB is a standard 1.6-mm, double-sided FR4 substrate. Figure 2.2 indicates that the width of a 50-ohm track is a little less than 3 mm. To build the inductors, you need to use a thinner track with any arbitrary

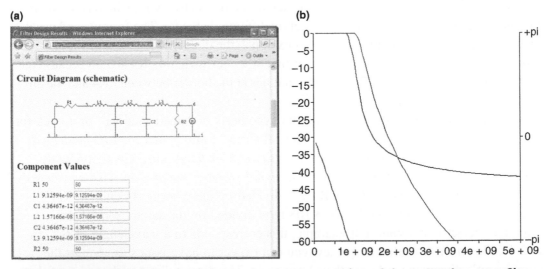

Figure 2.5: I designed the classic lumped component version of the 1-GHz low-pass filter with an Internet-based calculator. It used three inductors and two capacitors.

but convenient track width, for example, 100 ohms, which corresponds to a width of 0.678 mm. Then you just have to calculate the corresponding lengths, thanks to the equations in Figure 2.4. With this 100-ohm track, the 9.12-nH value can be achieved with a track length of 14.62 mm, and 15.7 nH corresponds to 25.16 mm.

Similarly for the capacitors, you must select an arbitrarily small impedance value, for example, 15 ohms, which corresponds to a track width of 15.2 mm, and the calculated track length for 4.36 pF is 10.48 mm. The final mandatory phase is to check that the largest dimension of the components is reasonably smaller than the wavelength of the highest frequency you are working with. Here, the largest length is the capacitor length, which is around 2.5 cm. You already calculated that 2.4 GHz was safe up to dimensions of 1 cm, so you can expect the filter to work correctly at 1 GHz, but it may start to be a little far from the predictions at frequencies of 2 to 3 GHz and higher.

The next step is to simulate the filter design. With RF designs, a simulation phase is always less expensive than some tens of PCBs thrown into the garbage bin. You have a couple of options for the simulator. The simplest solution is to use a circuit-only simulator, which has provisions for microstrip models, such as the free, useful Quite Universal Circuit Simulator (QUCS). It is efficient and easy, but the disadvantage is that you won't get a drawing of the physical microstrip design. At the other extreme, you can use a full-featured 2D or 2.5D electromagnetic simulator, such as Sonnet Software, which has a free Sonnet Lite version that can be used for designs as simple as this one. Results will be accurate, but the tool's complexity and calculation time are significantly higher. I like an intermediate approach: PUFF. This old DOS-based simulator is based on circuit models rather than on EM simulation, but it includes a somewhat graphical input of the microstrip design. Moreover, PUFF is nearly free because it comes with a couple of books (see References). It runs well in a DOS box under Windows XP, but, unfortunately, I wasn't able to make a screen copy, so you have only the screen shot in Figure 2.6.

The layout windows are a graphical representation of the filter design, with a 50-ohm track on both ends and a succession of thin, inductive, and wide capacitive segments. The simulation took just 5 s on my standard PC and shows good 1-GHz low-pass behavior but with a drastically reduced attenuation of around 3.5 GHz. A tool like PUFF can't explain why, but you already know the answer. At such a high frequency,

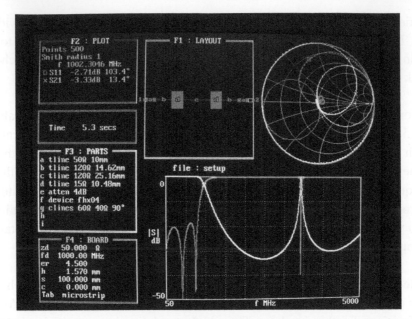

Figure 2.6: The PUFF application is a DOS-based microstrip drawing and circuit-model simulator. The bottom plot is the simulated transfer response and return coefficient of the filter. The behavior is well in line with the lumped version shown in Figure 2.4; however, a spurious response is expected at 3.5 GHz on the microstrip version.

the circuit dimensions are no longer "small" compared with the wavelength, and it's likely that a given track of the filter forms a tuned resonator at exactly 3.5 GHz. By the way, you can use a tool like Sonnet Lite to find out why by using the current density plot feature. A Sonnet design file is posted for your convenience on the companion website. Just play with it.

To the Bench

I couldn't resist the pleasure of building and testing this filter. Look at the PCB in Figure 2.7. You should be able to recognize the successive *L*, *C*, *L*, *C*, and *L* sections. I soldered two SMA connectors for the test, and, voila, the filter was ready. I hooked it to my old Hewlett-Packard HP8754A/H26 vectorial analyzer, which has a measurement range of 4 MHz to 2.6 GHz, and obtained the transmission measurement in Figure 2.8. Just compare it with the theoretical curve provided in Figure 2.5. Quite nice, isn't it?

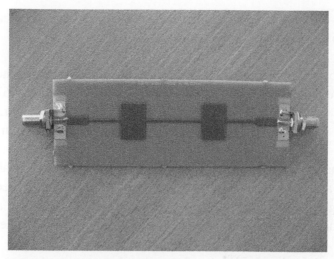

Figure 2.7: This is the assembled microstrip filter. From left to right you should recognize a small length of a 50-ohm track (connected to the SMA connector), an inductor made with 14.62 mm of thin wire, a capacitor made of 10.48 mm of wide track, then a longer inductor, another capacitor, and a final inductor. Funny, isn't it?

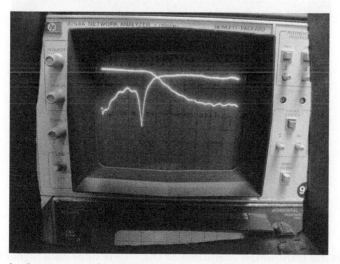

Figure 2.8: This is the measured response of the 1-GHz microstrip filter shown in Figure 2.5. The horizontal axis is from 8 MHz to 2.6 GHz. Curves are respectively transmitted and reflect coefficients with a 10-dB/division vertical scale. Impressively close to the simulation, with an attenuation of 20 dB from 1.5 to 2.6 GHz.

I measured the 3-dB cutoff frequency at 1030 MHz, close to the specification. The attenuation at 1.5 GHz is a reasonable 18 dB.

I had to switch to a different test setup to evaluate the performance of the filter at frequencies above 2.6 GHz. I used an even older Hewlett-Packard HP8620C microwave sweeper with a 2- to 8.4-GHz plug-in connected to the input of the filter, and a Hewlett-Packard HP8755 scalar analyzer on its output. The result I obtained is in Figure 2.9.

The first odd behavior was measured at 3.47 GHz with an unwanted peak response, which is exactly as expected through the PUFF simulation. At higher frequencies, the filter was no longer filtering anything. This was anticipated. Theoretical analysis showed that the filter dimensions were starting to be too large compared with the wavelengths.

I hope you are as pleased as I was when I saw that the actual performance was so close to the simulation—that is, at least up to 4 GHz! I must admit that this was the first time

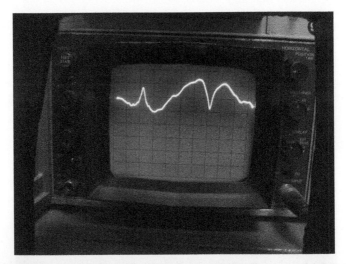

Figure 2.9: This is the transmitted coefficient of the 1-GHz filter on a higher frequency range. The horizontal scale is 2 to 8.4 GHz, the vertical scale is still 10 dB/division with 0 dB at the top of the screen. The isolation starts at 20 dB at 2 GHz but is significantly reduced at 3.47 GHz, in line with the PUFF simulation. The filter is unusable from 4 GHz upwards.

I saw actual results so close to the simulation. I was encouraged by this experiment, so I checked another aspect of microstrip circuits with the test PCB shown in Figure 2.10.

Textbooks explain that a 90° turn on a microstrip track must have a precisely calculated chamfered corner to keep the impedance under control. This is understandable. If you design a nonchamfered 90° turn, there is some "extra" copper on the corner. This copper acts as a small capacitor, which impacts performance. My test PCB included two identical 50-ohm tracks on an FR4 substrate, one with chamfers and one without.

I tested it with the same two test setups as the 1-GHz filter and was a little disappointed. There was no visible difference between the two designs up to 4 GHz. However, I then tested them at a high frequency, thanks to a 12.4- to 18-GHz plug-in I bought years ago for my HP8620C. I admit that trying to use an FR4 PCB at 18 GHz is a little risky, but it enabled me to see an impressive difference between the two versions. Refer to Figure 2.11 to see why you should always use chamfered corners on microstrips.

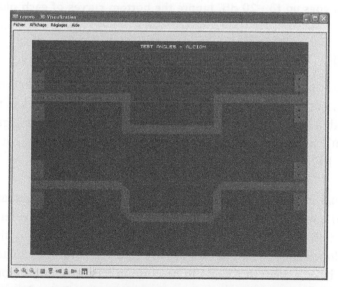

Figure 2.10: I designed this small PCB to check if chamfered corners on microstrips actually make a difference, as stated by the books.

(a)   (b)

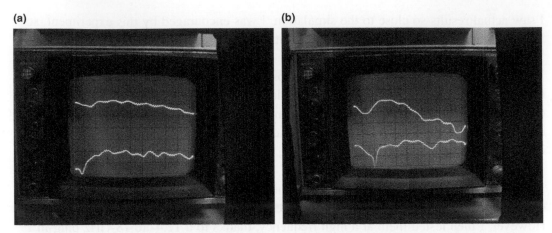

Figure 2.11: The top curves of each plot show the measured transfer coefficient of the respective chamfered (a) and unchamfered (b) microstrip lines at high frequencies, from 12.4 to 18 GHz. The vertical scale is 5 dB/division with the zero dB reference set at one division below the top. The difference was not visible at lower frequencies, but here it is impressive. With the chamfered version, the loss is around 5 dB, slowly increasing to 10 dB at 18 GHz. However, with the unchamfered version, the loss is far more erratic and goes up to −20 dB. Ignore the bottom curves of each plot.

Wrapping Up

Obviously, microstrip designs are unnecessary for a standard low-speed microcontroller design. However, knowledge of PCB track impedance-matching rules is mandatory when dealing with a large number of high-speed digital boards. Your gigahertz-clocked PC probably won't work without impedance-matched tracks. And, of course, their use is mandatory for RF designs.

"At which minimal frequency should I start to worry about track impedance matching?" That's a common question. The easiest answer is to compare the wavelength of the signal and the length of the longest track on your PCB. If the wavelength is far longer than your PCB dimensions, then you can usually safely ignore track matching. However, if you have a track that starts to be close to the wavelength you are working on, then you'd better take care and use impedance matching. Take the example of a classic 10-cm track on FR4. If you consider a factor of 10 "safe," then you should use matched tracks as soon as the wavelength on FR4 is shorter than 1 m (i.e., 10×10 cm), which corresponds to a wavelength in the air of 1 m times the square root of the 3.38 effective dielectric constant. The result is 1.84 m, which corresponds to a frequency of 300,000,000/1.84 = 163 MHz.

Do you use signals faster than about 150 MHz? If so, I hope that microstrips are no longer on the darker side for you.

itnched track's as soon as the wavelength on FR4 is shorter than 1 m (i.e., 10×10 cm) which corresponds to a wavelength in the air 1.1 m times the square root of the 5.38 effective dielectric constant. The result is 1.84 m, which corresponds to a frequency of $300,000,000 \div 1.84 = 163$ MHz.

Do you use signals faster than about 150 MHz? If so, I hope that microstrips are no longer on the dark side for you.

CHAPTER 3

Time Domain Reflectometry

In the previous chapters I talked a lot about impedance matching. In this chapter I will present the same subject from another angle: time domain. You may have read sentences such as "When matching is not perfect, a part of the signal is reflected back to the source." This may seem strange for engineers who are not used to high-frequency effects. Imagine the worst case of an impedance mismatch: a wire grounded at one of its ends. Do you think there could be any signal, reflected or not, in such a wire? Of course, and I will show it to you!

Signal reflection is in fact at the heart of an old but interesting measurement technique: time domain reflectometry (TDR). TDR enables you to detect, measure, and locate any impedance mismatch in a transmission line. In this chapter, I'll explain the theory. More important, I'll present some practical experiments to demystify these techniques. You will just need a good oscilloscope.

TDR Basics

Nothing can go quicker than $c = 3 \times 10^8$ m/s, the speed of light in free space (except guys jumping from black hole to black hole, if you trust some science fiction authors). The speed of an electrical signal going through a wire is a little slower than c because of the velocity factor of the transmission line, which is always slightly below unity.

Note: This chapter is a reprint of the article "The darker side: Time domain reflectometry," *Circuit Cellar*, no. 225, April 2009.

DOI: 10.1016/C2009-0-20196-6

Imagine that you have an infinite wire or a sufficiently long one terminated in its proper impedance-matching load, which is equivalent. Any signal will flow through the wire and be absorbed by its matched load. No problem, no reflection. Now imagine you have a long, perfect cable that is grounded at its far end. On the other end of the cable, connect a voltmeter and a current-limited 10-V power supply, and switch on the power. What will happen?

If you don't consider the cable length, then of course the power supply will be short-circuited to ground through the cable and the voltmeter will simply read 0 V. But there is no way to immediately know that the other end is grounded. The electrical signal will need to propagate through the cable up to the end to "see" that it is grounded. Then some information will need to return to give 0 V on the voltmeter. Practically speaking, if you replace the voltmeter with a fast oscilloscope, you will see that the line voltage is at first 10 V. It drops down to 0 V only $2T$ later, with T being the time needed for the electricity to travel through the wire! You can also interpret this phenomenon as the 10-V input signal being reflected back from the grounded end as a −10-V signal, giving 0 V as soon as both signals are summed up, and this is effectively the case.

In more complex applications, there may be several impedance changes through the wire, and each will reflect back a signal. *The shape of the reflected signal will be characteristic of the mismatch.* Its time position, relative to the initial pulse, will be directly proportional to the distance from the source. This is TDR, which is an invaluable technique for locating faults (e.g., in underwater communication lines and similar applications) and pinpointing impedance-matching issues (e.g., on high-speed PCB tracks). TDR can be performed either with a step signal as an excitation, as in my previous example, or with a quick pulse. I will use the latter in this chapter because the interpretation of the signals is a little simpler.

The basic setup for a pulse-based TDR system is shown in Figure 3.1. A generator provides a sharp and short pulse, which is sent to the transmission line to be tested through a signal splitter, enabling you to connect a high-speed oscilloscope without perturbing the impedance of the wire. The oscilloscope will then display both the initial pulse and any pulses reflected by the wire. Note that the length of the cable between the splitter and the oscilloscope doesn't matter because both the initial pulse and the reflected pulses have to support the same delay through this cable.

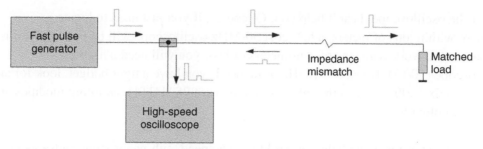

Figure 3.1: A time domain reflectometer includes a fast pulse generator and a way to display the reflected pulses, usually a high-speed oscilloscope. Thanks to a signal coupler, the oscilloscope enables you to display both the initial pulse and any reflected signals.

Let me write a few words about 50-ohm signal splitters. Such a splitter can be built with three 17-ohm resistors in a star configuration. The 17-ohm value enables you to keep a 50-ohm impedance on all branches. Why? Because each of the two output branches is supposed to be connected to a 50-ohm load, so each will have a 67-ohm impedance (i.e., 17 + 50 ohms) thanks to the 17-ohm serial resistance. This gives 33.5 ohms as both branches are in parallel. Just add the last 17-ohm resistor in the series and you are back to 50 ohms. Magical, isn't it? So you can build a 50-ohm splitter with just three resistors, but it is far easier to achieve good performance with an off-the-shelf splitter, especially when manipulating sub-nanosecond signals. The only disadvantage of such a resistive splitter is that a 6-dB loss is incurred in each of the two branches, but that's life.

1-ns Pulse Generator

Unfortunately, there is a problem with TDR techniques. If you need a good distance resolution, you must generate and detect very short pulses. Consider a standard transmission line with a velocity factor of, for example, 0.8. The speed of light is 30 cm/ns in free space, so it is 24 cm/ns (i.e., 0.8 × 30 cm/ns) in this wire. Because the signal must go back and forth, you will need to be able to manage 1-ns signals to get a 12-cm distance resolution. There are two issues: oscilloscope and generator.

As for the oscilloscope, I can't help you. Of course, if you just need to locate a problem within tens of meters, a low-cost 50-MHz oscilloscope will be fine. However, if you need to work with tens of centimeters or less, you will need a high-end oscilloscope (500 MHz or even 1 GHz or more). If you have a tight budget, look for an old Tektronix 7000 series on the Internet, maybe with 7S11/7T11 sampling modules and an S-4 sampling head.

As for the generator, I can help you build a high-speed, sub-nanosecond pulse generator for less than $5. As you can see in Figure 3.2, it can't be simpler, right?

Well, I must admit that I had to read it twice when I first saw this concept in an old National Semiconductor application note (see References). It is quite unusual to see a transistor with a grounded base generating anything—particularly ultra-high-speed pulses. K1 is simply a DC/AC high-voltage converter—the kind of converter used for CFL display backlights. With D1 and C1, this is a convenient, inexpensive way to generate a 300-V DC voltage. Yes, 300 V. This voltage is used to charge the small C2 capacitor (1.5 pF) through R1, and this is where the magic happens. At a given point in time, the voltage on C2 exceeds the avalanche breakdown voltage of Q1, usually around 60 to 80 V. Q1 then briefly conducts and discharges C2 through R3. This generates a pulse on the output; the pulse's duration will be roughly proportional to C2. More important, the pulse's rise time will be short because the avalanche phenomenon is fast due to the underlying physics. Intuitively, with such a high voltage the electrons will have a lot of energy and will be able to jump over the transistor's barrier very quickly. Some transistors are better than others for this application. The old 2N2369 is fine, so I have one. If you need longer pulses, just increase C2.

Where Are Your Magnifying Glasses?

Figure 3.2 is simple, but you need to be careful as you build it. Its performance will depend on the parasitic component values. Any useless wire in the critical section of the design (i.e., between C2, Q1, R3, and the output connector) will inevitably introduce parasitic inductances, which will drastically degrade the pulse generator's performance. You need to build it as small as possible. Surface-mount versions for C2 and R3 will at least yield better results than classic packages; however, I don't know if there are good SMT equivalents for the 2N2369.

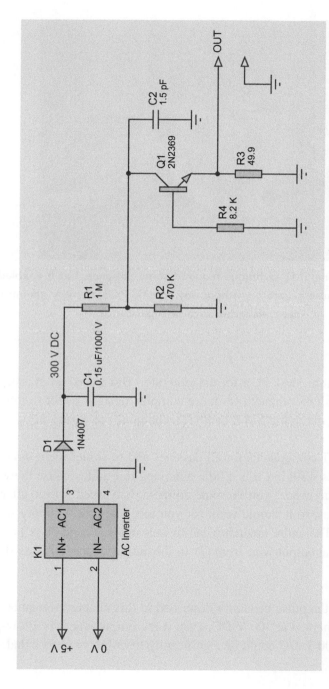

Figure 3.2: You can build a sub-nanosecond pulse generator for $5 or less using an avalanche mode generator. A high-voltage generator, here built using a CFL backlight power supply, drives an NPN bipolar transistor in its avalanche region, which generates a fast pulse on the output.

Figure 3.3: The flying SMT technique required some patience, but it enabled me to build a compact pulse generator without any specific PCB, thereby minimizing any parasitic inductance or capacitance.

You may design a custom SMT PCB for the generator. But I used an unusual assembly technique, which I don't recommend for more complex designs or for trembling engineers. Let's call it the flying SMT, or FST, technique (see Figure 3.3).

The idea is to use SMT components for all passives and to solder them directly to the 2N2369 leads. It works well, but it is a little annoying to build because these nasty SMTs don't want to stay where you've soldered them. It worked for me after soldering patiently for 30 minutes, so it should work for you too. The power supply section of the design is not critical. The entire assembly can fit in a small shielded box (see Figure 3.4). Just make sure the output wire from Q1 to the output connector is as short as possible.

Caution! If you build the pulse generator described in this chapter, remember that there is a capacitor inside charged at 300 V DC, even if the output signal is a low-power 50-ohm signal. The 300 V DC could be significantly harmful, or even lethal, so please take care.

Figure 3.4: I soldered the pulse generator transistor directly on the output connector and added a reused CFL backlight DC/AC converter with a 1N4007 rectifying diode and a small 1000-V ballast capacitor to provide a 300-V DC supply. In fact, 100 V would be enough. I added a small heatsink on the transistor just in case, but it seemed useless.

First Experiments

I am sure you want to know about the actual performance of this $5 avalanche generator. Figure 3.5 was taken with a high-end 1-GHz Lecroy WaveRunner 6100 digital oscilloscope (which provides no less than 10 Gsps single-shot and 200 Gsps equivalent sampling speed for repetitive signals) using 50-ohm input impedance.

The pulse rise time was measured at 244 ps, including the rise time of the oscilloscope itself, which is specified at 400 ps, so the pulse generator may be far quicker! The pulse width is around 0.5 ns, which is not bad. Such a pulse has frequency components of up to 1 GHz or so, which means it could be a helpful generator for numerous experiments. Keep it in on your bench just in case.

It is time to show you my first actual TDR measurement. The test setup is in Figure 3.6. The pulse generator is connected to an off-the-shelf Greenpar three-way 50-ohm

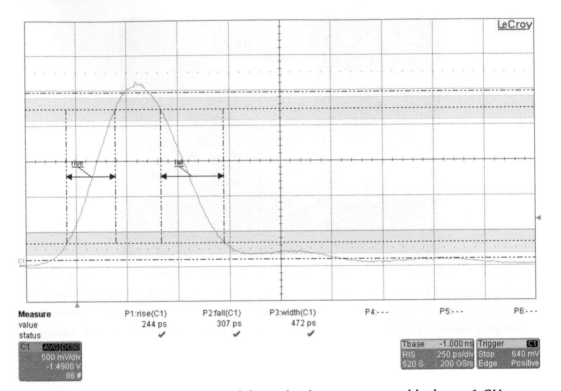

Figure 3.5: This is the output of the avalanche generator, grabbed on a 1-GHz digital oscilloscope. The pulse rise time is measured at 244 ps, far below the oscilloscope's specified rise time. The pulse width is less than 0.5 ns.

resistive splitter (obsolete). One of the outputs of the splitter is connected to the oscilloscope through a 50-ohm cable. The other is connected to a 1.5-m open-ended SMA cable (see Figure 3.6). Switch on the oscilloscope and set the trigger voltage high enough to synchronize only on the initial pulse. You get the image in Figure 3.7.

As expected, there is a reflected pulse of 15.66 ns later than the original signal. Assuming a velocity factor of 0.7, and keeping in mind the factor of 2 for back and forth directions, this means the discontinuity was 1.64 m away (i.e., $0.7 \times 3 \times 108 \times 15.66.10–9/2$ m)—not too far from the actual 1.5-m cable length. The difference is

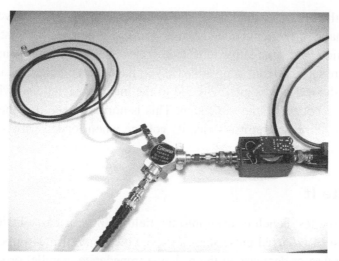

Figure 3.6: Check out my TDR setup. The custom pulse generator drives a three-way 50-ohm splitter (an old Greenpar model in this case). One output of the splitter (on the bottom) is connected to the oscilloscope through a 50-ohm coaxial cable. The other drives the system under test, which is a simple 1.5-m unterminated SMA cable.

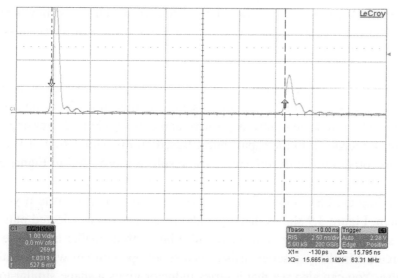

Figure 3.7: When a TDR is connected to an open-ended transmission line, there is a positive reflected pulse. The amplitude of the reflected pulse is equal to the incident pulse, but here the resistive three-way splitter induces a 6-dB loss. Thus, the voltage is theoretically divided by two (here a little more due to additional losses).

probably due to the delays in the splitter itself or to a slightly different velocity factor. Moreover, the shape of the reflection signal is useful. Remember my first example of a shorted wire, which gave a negative reflected signal? The line is open ended, and in such a case, the reflected signal is positive. If you have a step generator rather than a pulse generator, the reflected signal will add to the incident signal and will double its amplitude because both have the same sign. This is normal given that the voltage on an open-ended 50-ohm generator is twice its voltage when loaded with a matched 50-ohm load.

Let's Simulate It

Before going to the test bench or even into the field with your basic TDR system, it is nice to have a list of reflected pulse shapes for the different "usual" impedance mismatches: increase or decrease in the resistive impedance, parallel or series parasitic capacitor or inductance, and more. I could have built a dozen different test benches and measured the actual behavior, but simulation is a wonderful time-saving tool. The only issue is that a classic analogue linear simulator such as Spice can't easily handle line-length effects, so it isn't appropriate to simulate TDR effects. Fortunately, you can use the free Quite Universal Circuit Simulator (QUCS), which I used in Chapter 2. The QUCS simulation of a parasitic parallel capacitor in the middle of a transmission line is shown in Figure 3.8.

When a pulse is applied to a capacitor, this component first behaves as a short circuit, giving a negative reflected pulse similar to a short-circuited line. Then the capacitor slowly loads, and the reflected pulse is positive and exponentially decreasing, with a time constant proportional to the capacitance. Based on this simulation, it is easy to simulate all the other classic disturbances. The results are provided in Figure 3.9.

All corresponding QUCS simulation files are on the companion website if you want to play with them. Figure 3.9 shows that a parallel resistor gives a reflected pulse shape similar to a short-circuited line: a negative pulse but with a smaller amplitude than a full short circuit. Similarly, a series resistor is a small open circuit with a small positive reflected pulse. You can also see that a series inductor gives a shape similar to a parallel capacitor but with an opposite polarity. Parallel inductors and series capacitors also display dual behavior.

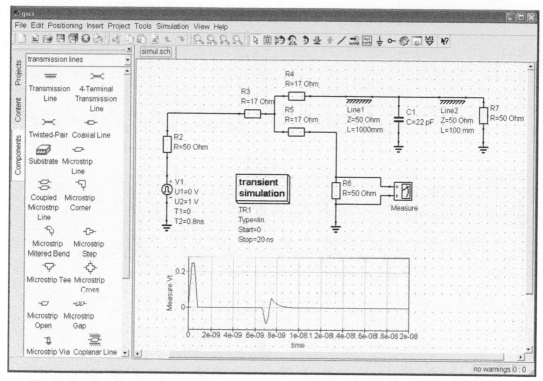

Figure 3.8: QUCS makes it easy to simulate a TDR experiment. A pulse generator drives an ideal three-way resistive signal splitter made with three 17-ohm resistors. One output drives a transmission line (here two 1-m lines with a parasitic parallel capacitor in the middle). The other drives a virtual voltage probe.

Real Life Versus Simulation

Let's compare simulation with real life. I'll begin with a parallel capacitor. Note that I reused the small S-shaped 50-ohm microstrip PCB that was built for Chapter 2. I soldered a 22-pF 0805 SMT capacitor in the middle of the microstrip line, with its other end grounded (see Figure 3.10).

Next I connected the microstrip board at the end of the SMA, cable used in Figure 3.6, connected another 1.5-m cable at the other end of the microstrip test board and finally

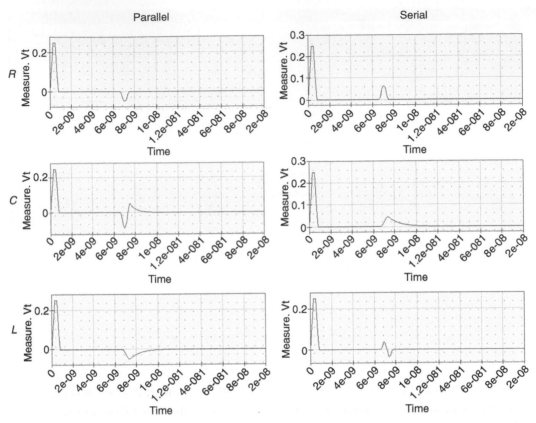

Figure 3.9: These are the results of the QUCS simulation of the reflected shapes for the six elementary impedance mismatches: parallel and serial *R*, *C*, and *L*.

used a 50-ohm SMA load to provide proper matching. So the test setup is a 3-m, 50-ohm line with a 50-ohm load at its end but with a parasitic 22-pF parallel capacitor to ground at the middle. I switched on the oscilloscope and pulse generator, and, voila, I got what you see in Figure 3.11a.

Comparing it with the theoretical shape for parallel capacitors, Figure 3.9 shows that the overall shape is similar, with first a negative pulse and then a smaller positive one. There are other small pops, probably due to other impedance mismatches (i.e., far from perfect ground connection of the capacitor, non-ideal capacitor, non-ideal connectors, etc.). But once again, the overall shape is similar.

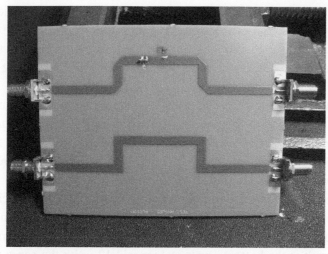

Figure 3.10: I reused the microstrip test board I presented in Chapter 2. I added
a parasitic serial or parallel 22-pF SMT 0805 capacitor in the middle of the line
to compare the theoretical TDR behavior and an actual one.

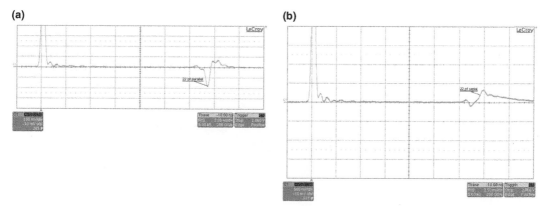

Figure 3.11: Experimental TDR waveforms with (a) a 22-pF parallel capacitor or
(b) a22-pF series capacitor in the middle on a microstrip transmission line. Just
compare these shapes with the corresponding theoretical shapes provided
in Figure 3.9.

Do you want another test? This time I hooked the same capacitor in series with the line, just by cutting 1 mm or so out of the microstrip and soldering the 22-pF capacitor across the gap. The result is Figure 3.11b. It is similar to the theoretical shape for a series capacitor but with additional bumps, in particular at the beginning of the pulse.

Another interesting use of TDR techniques is in evaluating the performance of connectors, particularly at high frequencies. TDR will easily show you any defective or less-than-ideal connectors. I performed a simple test by hooking a good 50-ohm SMA load at the end of a test 50-ohm cable. Theoretically, a 50-ohm load shouldn't reflect anything. But because the load and the SMA connectors were not 100% perfect, there was a small reflected pulse. I increased the vertical sensitivity of the oscilloscope and was able to see it easily. Refer to the top curve in Figure 3.12. Next, I inserted an SMA-to-BNC and BNC-to-SMA adapter pair between the SMA cable and the same SMA load. The result is the bottom curve in Figure 3.12, with the same vertical settings.

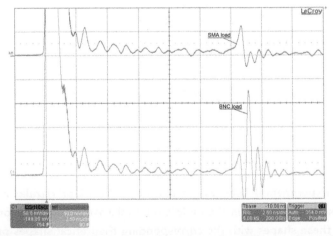

Figure 3.12: The vertical scale is increased to show the difference between the parasitic reflections with an SMA 50-ohm load (*top*) and with the same load connected through two SMA/BNC adapters (*bottom*). Conclusion: I don't like BNC.

Do you see a difference? I conclude that you shouldn't use BNC connectors for high-frequency designs if you're looking for good impedance matching.

Wrapping Up

I covered some of the potential applications for TDR. Even with a poor man's pulse generator and a good oscilloscope, you can easily pinpoint impedance-matching problems on cables and transmission lines. Moreover, a quick look at the shape of the reflected pulse will enable you to get a qualitatively good idea of the kind of defect. I read that TDR engineers working on the maintenance of submarine lines can easily guess if a problem is related to water ingress, corroded contacts or something similar just by looking at the TDR shapes. You now know why.

I haven't discussed the mathematical aspects of TDR. This would be the subject of another interesting chapter, particularly because a simple fast Fourier transform of the TDR signal can bring you back into the frequency domain. You can then deduce the line's bandpass simply by looking at its TDR shape, at least if there aren't any losses. As a TDR setup works only by looking at reflections, it may not detect a signal that is absorbed in a line and not reflected back.

I hope this journey in TDR has been enjoyable. Don't hesitate to test it, play with it, and send me the results of your experiments. TDR is no longer on the darker side for you.

Part 2
Electromagnetic Compatibility

CHAPTER 4

Let's Play with EMI

We will now switch to another black magic domain. It is something that can give headaches to engineering managers evaluating the risks on their projects. It is something that is often not learned until it is too late. It is electromagnetic interference, or EMI for short.

EMC Basics

Electromagnetic compatibility (EMC) is about being reasonably sure that a given piece of equipment will work peacefully with other equipment in its neighborhood. "Peacefully" means the equipment should not be perturbed by others (EMI immunity) and it should not perturb them (EMI emission limitation). Let's look at the details.

In the EMC/EMI world, equipment is usually classified into two categories: voluntary versus unintentional transmitters (or receivers). For example, a voluntary transmitter could be your remote control transmitter, while an unintentional transmitter could be a badly designed power supply generating plenty of EMI noise. Similarly, a voluntary receiver could be your FM radio, while an unintentional receiver could be your MP3 player picking up RF noise transmitted by your cell phone. Now we have enough vocabulary. EMI immunity relates to voluntary transmitters (generating strong RF fields because they are designed to transmit something) and unintentional receivers

Note: This chapter is a corrected reprint of the article "The darker side: Let's play with EMI," *Circuit Cellar*, no. 205, August 2007.

© Elsevier Inc.
DOI: 10.1016/C2009-0-20196-6

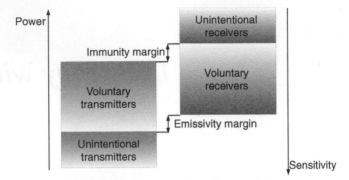

Figure 4.1: This figure illustrates the two main EMC requirements: EMI immunity, which is related to the interaction between voluntary transmitters and unintentional receivers, and EMI emission limitations, which are related to the interaction between unintentional transmitters and voluntary receivers.

(low-sensitivity receivers) that could be agitated by these strong RF fields. Reciprocally, the EMI emission limitation is a concern between unintentional transmitters (low-level but not expected) and voluntary receivers (high-sensitivity because they are designed to receive). Inspired by Tim Williams's book, *EMC for Product Designers*, Figure 4.1 illustrates these two fundamental EMC gaps.

Because nobody knows which EMI environment a given product will be used in, legislation defined the absolute maximum admitted emission power and minimum immunity requirements. Legislation on EMC was initially limited to radio communications (i.e., voluntary transmitters and receivers), but it was extended to unintentional transmitters about 20 years ago, particularly in Europe. The first European Commission EMC directive, 89/336/EEC, was ratified in the early 1990s and was made mandatory in 1996. It covers both immunity and emissions limitations. In the United States, FCC Part 15 limits emissions of unintentional transmitters (spurious emissions, subpart B) as well as low-power, unlicensed transmitters (subparts C to G). But as far as I know, emission compliance with RF immunity standards is not mandatory for most products under the FCC's guidelines, even if additional standards are usually requested by the marketplace. In all other countries, the international standards IEC61000-xxx from the IEC's TC77 are usually translated into national standards, providing a similar framework.

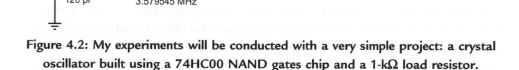

Series resonant test oscillator

Figure 4.2: My experiments will be conducted with a very simple project: a crystal oscillator built using a 74HC00 NAND gates chip and a 1-kΩ load resistor.

Finally, electromagnetic signals can be either radiated (through the air) or conducted (through wires). Standards usually deal with these two aspects separately, even if they are often closely linked, as you will see.

Okay, you probably think that I was lying in my introduction when I promised to focus on application-oriented explanations and pragmatic tips rather than on long and painful theory. Agreed—I am going to the bench. Because the EMC subject could fill up thousands of pages, I will concentrate on only one topic: unintentional emissions. To demonstrate the basics of this nasty phenomenon, I will not describe a rocket science project. I'll focus on one of the simplest unintentional transmitters: a small TTL oscillator built around a 74HC00 chip and the 3.5795-MHz crystal I had on my desk, with a 1-kohm load resistor simulating a clocked device on the board (see Figure 4.2).

Before I demonstrate the phenomenon, I need to introduce the EMC guy's "pocket knife."

The Spectrum Analyzer

You will need a way to see and measure RF because EMI is mainly about RF signals covering a large frequency range (from nearly DC to several gigahertz) and a large signal amplitude range (from watts down to nanowatts or lower). A spectrum analyzer, an expensive but invaluable tool, must be used. Oscilloscopes are currently available far above the gigahertz region (even up to 18 GHz for those with really deep pockets), and high-end digital models provide a spectrum analyzer using a fast Fourier transform of the time-acquired signal. However, an RF analyzer is usually far more accessible from the used market. A spectrum analyzer is nothing more than a tuned, swept, very selective, and very sensitive receiver. The worst real spectrum analyzer provides at least 80 or even 100 dB of dynamic range. The architecture of a simple RF super-heterodyne spectrum analyzer can be seen in Figure 4.3.

The measured signal, coming from a wire or an antenna, first goes through a step attenuator to avoid saturation. A low-pass filter gives it a clean signal from kilohertz up to the maximum frequency of the equipment. Assume that the maximum is 3 GHz. The signal is then mixed with a clean high-frequency swept local oscillator covering a 3-GHz range (from 3.5 to 6.5 GHz, for example). It is bandpass filtered around a frequency just higher than the maximum usable frequency (here 3.5 GHz). This architecture allows it to limit spurious receptions. For example, when the local oscillator is set at 4.223 GHz, the only input frequency that will provide a signal going through the 3.5-GHz filter is at 723 MHz (i.e., 4.223 GHz − 3.5 GHz). The 3.5-GHz intermediate frequency signal is amplified and down-converted through several mixing stages to a low intermediate frequency. Then one of the most important parts of the spectrum analyzer comes in: the resolution filter.

This is a very selective, calibrated, and variable bandpass filter that allows the analyzer to set the measurement bandwidth, which has a strong influence on sensitivity, frequency discrimination, and measurement speed. The signal is sent through a logarithmic amplifier, allowing the analyzer to display a power level in a logarithmic scale over several decades. Finally, the signal's amplitude is detected, filtered again to remove noise by averaging, and displayed.

This presentation is very simplified because digital techniques are heavily used, from the latest intermediate frequency blocks down to detection through digital resolution filters and FFT. Moreover, the microwave equipment that has a frequency range far above a few gigahertz uses a different conversion scheme for its input section. It uses exotic Yig-based preselection bandpass filters and harmonic mixers. Anyway, this simplified architecture fairly matches the architecture of 20-year-old spectrum analyzers, such as my

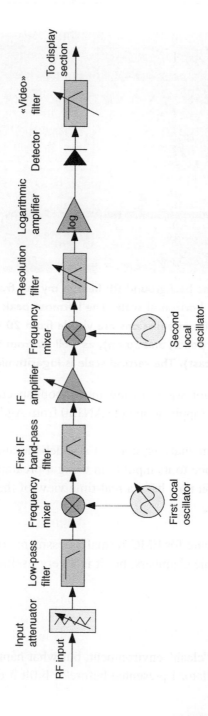

Figure 4.3: Here is the overall, but largely simplified, architecture of a super-heterodyne low-band RF spectrum analyzer.

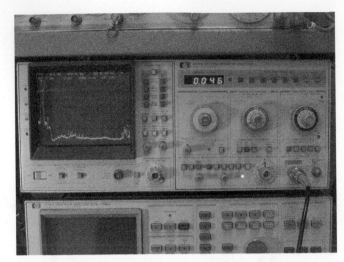

Figure 4.4: The plot shows the background RF field in my lab from nearly DC up to 100 MHz with a 10-MHz/division horizontal scale. The leftmost peak at 0 Hz is an artifact of the spectrum analyzer. Significant emissions are visible from 20 to 50 MHz with a peak at 40 MHz (short waves, police, and so on), as well as from 90 to 100 MHz (FM broadcast). The vertical scale is logarithmic.

Hewlett-Packard HP8569. If you want to know more about spectrum analyzers, the bible is "Spectrum analysis basics" (application note AN150 from Agilent).

Now that you have a spectrum analyzer, connect a wideband antenna or, better yet, a preamplified electric field probe to its input (I'm using a preamplified probe from a Hameg HZ530 probe set). You now have a real-time view of the surrounding RF environment (see Figure 4.4).

Such a test setup is not adequate for EMC formal assessments requiring calibrated antennas and shielded anechoic chambers, but it is a very useful relative measurement tool, as you will see.

Radiated EMI

Figure 4.4 was taken with a "clean" environment, but what happens when you switch on the small test crystal oscillator I presented before? I built it on a small prototyping

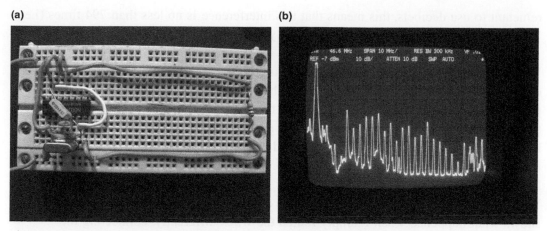

Figure 4.5: (a) The test crystal oscillator in its first nonoptimized version. (b) The spectrum received 2 m away from the test oscillator in (a). The spurious emissions are nearly 1000 times more powerful than the FM broadcast signals!

board, even if you don't agree that it is the nicest idea for a good RF design (Figure 4.5a).

I then dropped the prototype on a desk 2 m away from the receiving antenna (that's around 6 or 7 feet for those living on the wrong side of the ocean), and powered it from a 5-V bench supply. Figure 4.5b shows what was displayed on the spectrum analyzer's screen. I am not joking! The separation between peaks is close to 3.5795 MHz, which is not a surprise. You can see all the harmonics of your crystal oscillator from 20 MHz up to a few gigahertz.

That's a good illustration of unintentional emissions, isn't it? Note the power of the maximum peak on this first test; we will use it as a reference. The top of the screen is −7 dBm. The vertical scale is 10 dB/division. I found the maximum peak at about −34 dBm (including a gain of the preamplified antenna, air path loss through the 2 m between the test oscillator and the antenna, and so on). Go back to the reference spectrum in Figure 4.4. In the same setup, the FM broadcast peaks were measured at −63 dBm. That means that the EMI field transmitted by the small crystal oscillator is 29 dB (i.e., 63 dBm to 34 dBm) more powerful than the FM broadcast signals, even if the crystal oscillator is 2 m away from the receiving antenna. And, if you are still reluctant to use decibels, this means that this interference is no less than 794 times (i.e., $10^{29/10}$) more powerful than the radio station you want to listen to!

What happened? Any AC current going through a conducting loop generates an electromagnetic field that will propagate in the space with an amplitude roughly proportional to the current, to the surface of the loop, and to the square of the frequency. Because your crystal oscillator is more or less using square waves, all harmonics of the clock signal are transmitted and, unfortunately, well received up to the maximum frequency of the HC00 gates.

What should you do? First, try to reduce the area of the current loops. Figure 4.6a–b shows what happened when I simply routed the wires differently, connecting the 1-kohm resistor with two closely spaced wires.

The spectrum looks the same, but the maximum peak is now –39 dBm. That seems like a small improvement, but remember that you are using a logarithmic wide dynamic scale. That's a 5-dB improvement (from –34 dBm down to –39 dBm). It means that the power of the received signal is $10^{5/10}$ lower, so it was reduced by a factor of 3.2 just by moving a wire. Not so bad. *The golden rule is to limit all loops carrying high-frequency signals or low-frequency signals with abrupt slopes such as logic signals.*

(a)   **(b)**

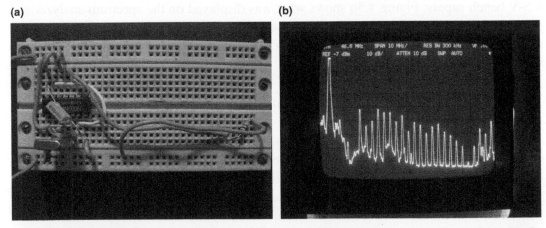

Figure 4.6: (a) The first improvement to the test oscillator was to move the resistor load wire to reduce the loop area. (b) Compared with Figure 4.5b, the improvement in this photo provides a reduction of 5 dB in the unintentional emission—small, but not bad.

Because the return-current path is usually the ground, this recommendation translates into the most important EMC tip: *Build your circuits with a full ground plane on one layer of your PCB*. That will ensure that there is always a return path just below every wire. Some will argue that split ground planes can be good, but I'm sure that 99.9% of designs will work better with a full ground plane rather than with more complex schemes (except for some audio designs well packed in a shielded enclosure and not really affected by radiated EMI). For example, if you design a board using two ground planes joined only by a small bridge, then there will be current loops, except if all other wires going from one area to the other are passing exactly over the bridge—which is not easy to do.

How do you go further with minimal changes to the design? The level of harmonics generated by our test project is as high as it is because the signals are steep square signals. But is this steepness actually needed? Why use a 74HC00 with a propagation delay of 7 ns, high output current, and, more important, heavy current peaks at each transition, due to the CMOS technology of the HC chip family, to build a simple 3.5-MHz clock? I dug into my shelves and swapped the HC00 with an old LS00. The maximum peak was immediately reduced to −45 dBm, which is a 6-dB incremental improvement. Compared with the initial situation, these two small changes reduced the transmitted noise by 11 dB, or more than a factor of 10, just by moving a wire and swapping a chip with an older one!

Let me summarize the lesson learned from this second experiment. *Always use the slowest possible technology for a given project*, with the slowest possible output currents and the slowest possible voltages, particularly in digital signals. That's obvious, but it's too late when you remember it after the design is complete.

Conducted EMI

I didn't want to talk about conducted EMI in this chapter, but I have to. Assume that the noise level is still higher than acceptable. You can continue to improve the design everywhere you can and see the result on the transmitted noise 2 m away, but that could be time-consuming. It would be easier if you could pinpoint the precise location of the transmitted noise source. Fortunately, this is also possible with a spectrum analyzer. The only additional tool needed is a magnetic field probe (H-field probe). The reason is that the H-field is decreasing as the cube of the distance, while the E-field is decreasing as its square. So using an H-field probe is far more effective in locating a transmitter at very short ranges because the H-field is decreasing very quickly when you leave the immediate surroundings of the emitter.

You can buy good, low-cost, preamplified H-field probes such as the Hameg HZ530 probe set that I'm using, but after buying the spectrum analyzer your pockets may be empty. Fortunately, a simple H-field probe can be easily built using a small 50 ohm shielded wire. Just strip the cable a few millimeters from the end, turn it to make a small loop, and solder the inner connector back to the shield braid. The end of the braid should not be connected to itself, in order to provide a gap in the shield. That's it. Now you have a good magnetic field probe.

Assuming that you have an H-field probe connected to the spectrum analyzer, move it around the design and, in particular, around the wires connecting it to the 5-V power supply (see Figure 4.7).

Bingo! You can see that the power cables are radiating a strong RF field, in particular in the 50- to 100-MHz range. The current peaks drawn by the logic chip show up as conducted current peaks on the power lines because of insufficient decoupling, even

(a)   (b)

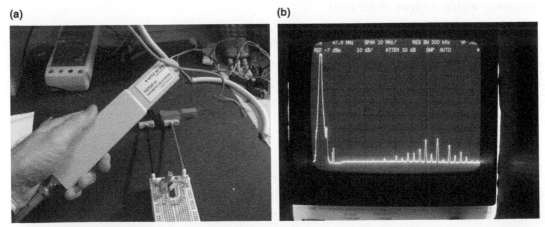

Figure 4.7: (a) Using an H-field probe allows the spectrum analyzer to pinpoint a key contributor to the radiated emissions, namely the 5-V and GND power wires. (b) The H-field probe shows a high level of radiated signals from the power lines.

(a)   **(b)**

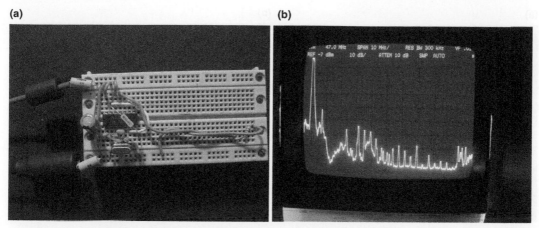

Figure 4.8: (a) This is the next improvement: coils and ferrite filters on the power lines and a slower LS00 chip. (b) With the improved design in (a), the unintentional emissions are drastically reduced.

though the ubiquitous 100-nF capacitor is in place on the power rails. The conducted EMI is then radiated because of the long unshielded power wires. That's why conducted and radiated EMI are often linked. You must reduce the conducted EMI through the power lines. I modified the test oscillator with the addition of a 1-mH coil plus a ferrite bead on both the GND and 5-V lines, as well as the addition of a 10-μF low-ESR capacitor in parallel to the 100-nF one (see Figure 4.8a).

One caution regarding filtering coils: At high frequencies, a higher-value coil is not always better because parasitic capacitances can quickly make it useless. Technology does matter, and that's why ferrite coils are usually far more efficient for EMI reduction than standard coils of the same value, but this subject would need a chapter by itself.

Back to the spectrum analyzer. The maximum received peak is now at only −51 dBm if you consider only the high-frequency (HF) contributors above 10 MHz, which is another 6-dB improvement! (See Figure 4.8b.)

As a last improvement, I tried moving the 1-kohm resistor closer to the chip (see Figure 4.9a), and the received EMI level went down to −55 dBm, which is 4 dB better (see Figure 4.9b).

(a)

(b)

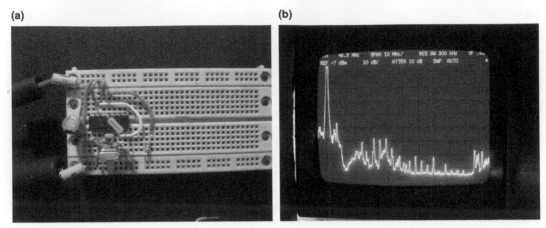

Figure 4.9: (a) This is the last improvement. The load resistor is moved close to the oscillator. (b) All of these improvements give an accumulated reduction in the unintentional emissions by 21 dB as compared with Figure 4.5b. That's more than a factor of 100!

Is That All?

Using one of the most simple designs (a crystal oscillator) built on the most simple platform (a protoboard), I showed you that a couple of very simple design modifications allowed me to reduce the unwanted emissions of the design by 21 dB (i.e., 55 dBm – 34 dBm), effectively reducing the spurious noise power by 99.2% (1021/10 = 125, 1/125 = 0.8%)! With these changes, the FM broadcast signals are now clearly visible on the spectrum analyzer, which means that you will be able to hear your favorite songs. And more important, these design changes were really minor, including rerouting some wires, an HC00 replaced by an LS00, coils, and ferrites on the power supply.

Impressive, isn't it? In this chapter, I have highlighted some key points of one aspect of EMC, unwanted radiated emissions, and some design tips that will make your life easier. The EMC-minded designer will also need to take care of immunity, conducted EMI, common mode noise, ESD, and the regulation side of the EMC. Anyway, I hope that I have successfully demonstrated that the EMC is not black magic, even if it is occasionally on the darker side.

CHAPTER 5

Cable-Shielding Experiments

I have in my technical library a book titled *Noise Reduction Techniques in Electronic Systems* by Henry W. Ott (AT&T Bell Labs). Although the book is nearly 40 years old, I am always amazed at its timeliness. It is the kind of book that any electronic design engineer must forever keep close by because the content is too fundamental and important to be outdated as technology changes. The book was first published in 1975 and then updated in 1988. In addition to chapters on grounding, filtering, shielding, ESD, thermal noise, PCB layout rules, and more, Ott devoted a chapter to cabling issues. What grabbed my attention was that he presented not only theories but also actual experiments on noise coupling with different types of cables and wiring setups. I thought this was definitely an interesting "dark side" subject, so I decided to try to reproduce the spirit of these experiments. I'll share my results with you.

Theory Refresher

Imagine that you have a wire conducting a low-level signal, say from a sensor, and another nearby wire conducting a noisy signal, say a digital signal. You know from experience that some noise will probably appear on your sensor signal. But how? As long as the signal and noise frequencies stay low, two main coupling mechanisms could be involved: capacitive coupling and inductive coupling. Capacitive coupling is the interaction of electric fields between the two wires. Inductive coupling is the interaction of magnetic fields between the two circuits. At higher frequencies, or at longer

Note: This chapter is a corrected reprint of the article "The darker side: Cable shielding experiments," *Circuit Cellar*, no. 219, October 2008.

© Elsevier Inc.
DOI: 10.1016/C2009-0-20196-6

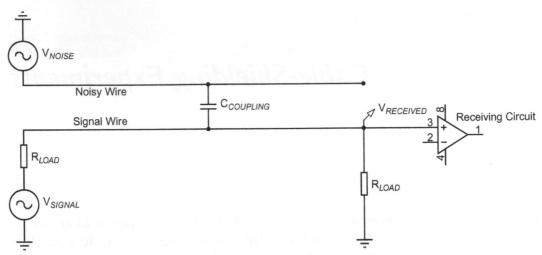

Figure 5.1: This is the equivalent circuit of a capacitive coupling between two wires. Capacitively coupled noise is a problem mainly with high-frequency noise, high impedances, and closely spaced wires.

distances, a combination of both effects can lead to electromagnetic coupling. But I won't cover this. I will focus here on short-distance effects.

Let's start with *capacitive coupling*. When two wires are close to each other, there is a parasitic capacitance between them, which can be depicted on an equivalent electric circuit (see Figure 5.1). The parasitic capacitance depends on the length, distance, and geometry of the wires. When associated with the loading impedance of your signal wire, the capacitance implements a high-pass RC filter between the noise source and your signal receptor because the capacitance is in series between the noise source and the receiver circuit.

The noise voltage added on $V_{RECEIVE}$ is calculated as

$$V_{noise\ received} = 2\pi F_{NOISE} C_{COUPLING} V_{NOISE} \frac{R_{LOAD}}{2}$$

Therefore, the noise voltage added to your precious signal will be proportional to the noise source voltage, so the more noise you have in the nearby wire, the more noise you get. But it also will be proportional to the noise frequency; thus, capacitive coupling is mainly a problem of high-frequency noise. In addition, the noise voltage is proportional to the signal load impedance, which means that capacitance-coupling noise will be a major problem with high-impedance designs. Finally, it is proportional to the value of the parasitic capacitor: The higher the capacitance between the two wires, the more noise you get.

The good news is that this capacitive coupling is quickly reduced when the distance between the two wires is increased. The other good news is that using shielded cables theoretically removes any capacitive coupling as long as the shield is grounded at at least one point, except for the portion of the cable that will inevitably extend beyond the shield. So capacitive coupling is mainly a problem when two unshielded cables are close to each other, such as between wires of a ribbon cable.

Inductive coupling is a bit more nasty and difficult to imagine. When a current flows through a circuit, in this case the noise generator circuit, there is a corresponding magnetic flux generated that is proportional to the current going through the noise generator circuit. The magnetic flux will generate a noise current in any other nearby circuit (see Figure 5.2).

For inductive coupling the received noise voltage can be calculated as

$$V_{noise-received} = -L_{MUTUAL} \frac{dI}{dt}$$

I is the current going through the noise source loop, calculated as:

$$I = \frac{V_{NOISE}}{R_{LOADNOISE}} = \frac{V_{NOISEPEAK}}{R_{LOADNOISE}} \sin(2\pi F_{NOISE}t)$$

so $V_{noise-received}$ expands into

$$V_{noise-received} = \frac{2\pi L_{MUTUAL} V_{NOISEPEAK} F_{NOISE}}{R_{LOADNOISE}} \cos(2\pi F_{NOISE}t)$$

The noise voltage is thus proportional to the current going through the noise source loop ($V_{NOISEPEAK}$ over $R_{LOADNOISE}$), proportional to the noise frequency, and proportional to the mutual inductance between both circuits. The mutual inductance is roughly proportional

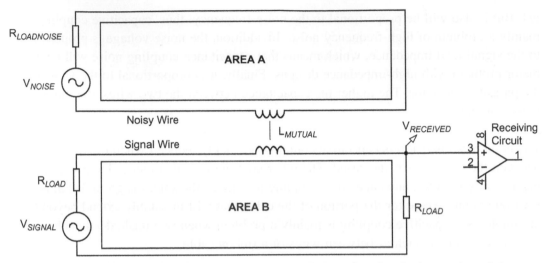

Figure 5.2: For inductive coupling, the noise is coupled on the signal wire as though it were a transformer. The coupling factor, or mutual inductance, is mainly dependent on areas A and B of the two-coupled circuits, their distance, and their mutual orientation.

to areas A and B of the two circuits, which are the areas between each signal and its corresponding return path. That's why the current return path should always be as close as possible to the signal path to reduce the inductive loop area.

Now the bad news. The coupling acts like a transformer with windings in series with both circuits, so the noise generated on the signal circuit is a voltage added in parallel to the wire. This means it is independent of the signal load impedance. In other words, when there is an inductive coupling, the generated noise voltage level is roughly the same for low-impedance or high-impedance signal circuits: R_{LOAD} isn't a factor in the formula. The contrary is true with capacitive coupling: This is even a way to distinguish between capacitive and inductive coupling. If you change the load impedance and the noise level stays roughly the same, you have inductive coupling.

The other bad news is that a shield may not reduce, even theoretically, an inductively coupled noise, especially if it is grounded only at one point. I will cover this topic in more detail later in this chapter. However, there is an efficient solution to inductive coupling: twisted wires. A twisted wire pair enables you to keep the net area between a conductor and its current return path at zero because the polarities are out of phase at each half-twist, at least if the currents are perfectly balanced.

Experimental Setup

I've covered the theory. Now I'll focus on the process of testing inductive coupling. I will also describe good shielding and wiring strategies to reduce noise coupling. It took me just 30 minutes to build the small test bench illustrated in Figure 5.3.

First, I needed to generate a strong magnetic field. I used an old Philips/Fluke PM5134A lab signal generator and tuned it to frequencies from 20 to 20 kHz. To

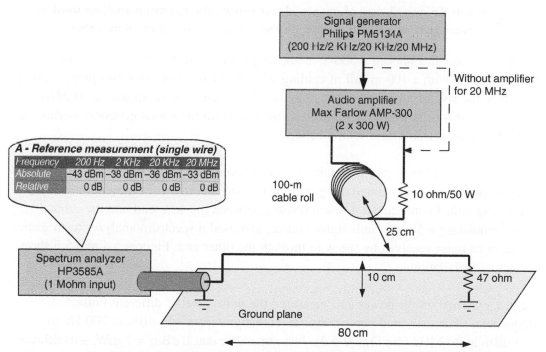

Figure 5.3: The reference configuration uses a single wire exposed to a strong magnetic field with a return path through a ground plane situated 10 cm below the wire. This configuration is our 0-dB reference by definition.

Figure 5.4: An overall view of my actual test bench. The spectrum analyzer used as a receiver is an HP3585 model, on the left at the end of the BNC cable.

amplify the signal, I hooked a 300-W audio amplifier and connected the speaker output to a coil made with a 100-m roll of cabling wire. I added a 10-ohm series power resistor to reduce the risk of blowing out the amplifier. I also used the setup with a 20-MHz frequency—this time with a direct drive to the coil from the signal generator because an audio amplifier doesn't have such a bandpass.

Then I used a 1.20-m × 50-cm metallic grid to build a reasonably good ground plane and placed a wire 10 cm above the ground plane and 25 cm away from the noise-generating coil. I connected a 47-ohm resistor between the wire end and the ground plane, emulating a 50-ohm null signal source, and used a spectrum analyzer to measure the level of noise received by the wire through the other end. Figures 5.4 and 5.5 show the actual installation.

Next, I switched on the power and measured the noise level at different noise frequencies. The measured noise power levels ranged from −43 dBm at 200 Hz to −33 dBm at 20 MHz (see Figure 5.3). Just remember that 0 dBm = 1 mW, −10 dBm = 100 μW, −20 dBm = 10 μW, and so on. The formula is

$$P(Watt) = 1\,\text{mW} \times 10^{dBm/10}$$

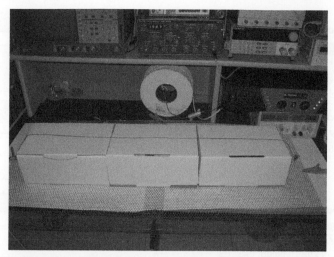

Figure 5.5: A close-up view of the experiment. The frequency generator is at the top and set to 1.9987 kHz. It is driving the 300-W power amplifier on the right (the one with LEDs). The amplifier output then excites the big coil made simply with a 100-m white wire roll. This coil generates a strong magnetic field, which is received by a simple wire. This victim wire is positioned 10 cm above a ground plane and connected to the ground, through a 47-ohm resistor on the right, and to the spectrum analyzer. The boxes are empty.

So the received levels were ranging from 0.05 µW to 0.5 µW. We'll use these values as a reference to compare other wiring schemes. These powers are "0 dB." Let's name this reference "configuration A."

Comparative Results

Imagine that this noise is far greater than what you can tolerate on your signal. Your first idea would probably be to use shielded cable, and that's what I've done. The results are illustrated in Figure 5.6.

If you replace the unshielded cable with a shielded one and connect only the shield to ground at one point, then the noise is reduced by only 2 dB! (See Figure 5.6, configuration B1.) If you think twice, that's normal because we are in inductive-

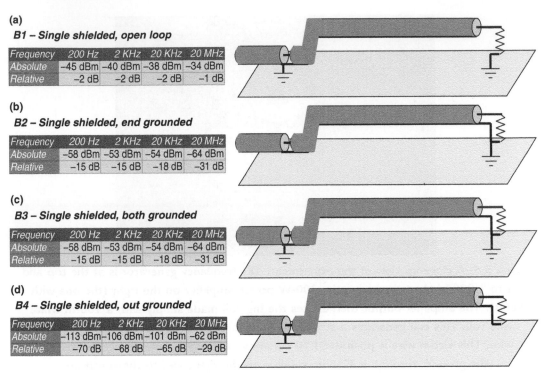

(a)

B1 – Single shielded, open loop

Frequency	200 Hz	2 KHz	20 KHz	20 MHz
Absolute	−45 dBm	−40 dBm	−38 dBm	−34 dBm
Relative	−2 dB	−2 dB	−2 dB	−1 dB

(b)

B2 – Single shielded, end grounded

Frequency	200 Hz	2 KHz	20 KHz	20 MHz
Absolute	−58 dBm	−53 dBm	−54 dBm	−64 dBm
Relative	−15 dB	−15 dB	−18 dB	−31 dB

(c)

B3 – Single shielded, both grounded

Frequency	200 Hz	2 KHz	20 KHz	20 MHz
Absolute	−58 dBm	−53 dBm	−54 dBm	−64 dBm
Relative	−15 dB	−15 dB	−18 dB	−31 dB

(d)

B4 – Single shielded, out grounded

Frequency	200 Hz	2 KHz	20 KHz	20 MHz
Absolute	−113 dBm	−106 dBm	−101 dBm	−62 dBm
Relative	−70 dB	−68 dB	−65 dB	−29 dB

Figure 5.6: Using a shielded wire doesn't improve the situation if the sensor emulation is still grounded (a) because the return path is still through the ground plane. However, an impressive improvement is achieved with a floating sensor (d). The two other configurations, (b) and (c), provide intermediate performance.

coupling mode and the shielded cable doesn't reduce the surface of the magnetic flow coupling area. Current can't flow through the shield. The return path is still 100% through the ground plane, which means it is 10 cm away from the cable.

But we can do better. Moving the ground point at the end of the cable close to the signal source reduces the noise by 15 dB, which is a significant improvement (see Figure 5.6, configuration B2). Grounding the cable at both ends doesn't change the result (see Figure 5.6, configuration B3); however, a huge improvement is possible with the same setup. If you connect the sensor ground to the cable shield (not to the ground plane), the noise is reduced by an impressive 65 to 70 dB, at least up to 20 kHz! (See

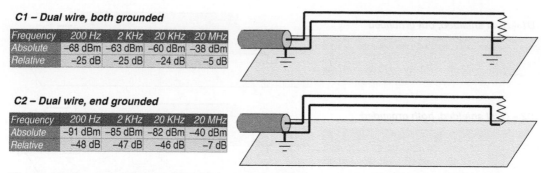

C1 – Dual wire, both grounded

Frequency	200 Hz	2 KHz	20 KHz	20 MHz
Absolute	–68 dBm	–63 dBm	–60 dBm	–38 dBm
Relative	–25 dB	–25 dB	–24 dB	–5 dB

C2 – Dual wire, end grounded

Frequency	200 Hz	2 KHz	20 KHz	20 MHz
Absolute	–91 dBm	–85 dBm	–82 dBm	–40 dBm
Relative	–48 dB	–47 dB	–46 dB	–7 dB

Figure 5.7: In comparison with Figure 5.6, a simple two-wire configuration already provides a reasonably good result at low frequencies but not in the higher frequency range.

Figure 5.6, configuration B4.) This is because this configuration forces all return current to flow through the shield close to the signal wire. However, at 20 MHz the improvement is less impressive: "only" 29 dB.

But were these improvements due to the shield or to the fact that the return path was far closer to the wire than in configuration A? To answer that question, I made a second test by replacing the shielded cable with a simple two-wire configuration (see Figure 5.7).

Strangely, the unshielded configuration (Figure 5.7, configuration C1) seems a little better than its shielded equivalent (see Figure 5.6, configuration B3) up to 20 kHz, but it is far worse at 20 MHz. Similarly, Figure 5.7, configuration C2, is significantly worse than Figure 5.6, configuration B4. In a nutshell, routing two wires close to each other drastically improves the situation (as compared with Figure 5.3), but a shielded cable nearly always performs better, especially at high frequencies. Not a surprise.

Can we do even better? You bet. One solution is to use a dual-wire shielded cable. The shield can be dedicated to a shield function, and the two inner wires can be used respectively for signal and return current. This configuration better avoids inductive coupling because the two wires are even closer and protected against electric fields by the shield. Figure 5.8 shows the corresponding configurations and measurements.

Once again, the best results are achieved when the shield is grounded only at the receiving side (see Figure 5.8, configuration D1). If you compare it with our previous

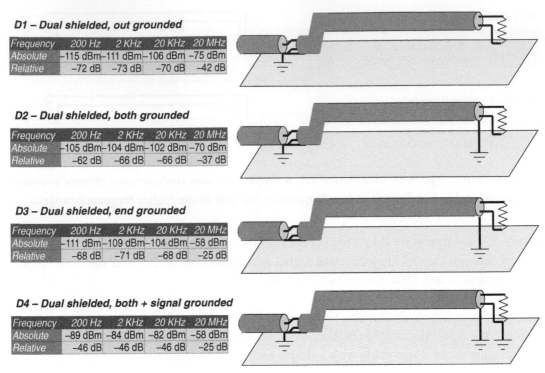

D1 – Dual shielded, out grounded

Frequency	200 Hz	2 KHz	20 KHz	20 MHz
Absolute	–115 dBm	–111 dBm	–106 dBm	–75 dBm
Relative	–72 dB	–73 dB	–70 dB	–42 dB

D2 – Dual shielded, both grounded

Frequency	200 Hz	2 KHz	20 KHz	20 MHz
Absolute	–105 dBm	–104 dBm	–102 dBm	–70 dBm
Relative	–62 dB	–66 dB	–66 dB	–37 dB

D3 – Dual shielded, end grounded

Frequency	200 Hz	2 KHz	20 KHz	20 MHz
Absolute	–111 dBm	–109 dBm	–104 dBm	–58 dBm
Relative	–68 dB	–71 dB	–68 dB	–25 dB

D4 – Dual shielded, both + signal grounded

Frequency	200 Hz	2 KHz	20 KHz	20 MHz
Absolute	–89 dBm	–84 dBm	–82 dBm	–58 dBm
Relative	–46 dB	–46 dB	–46 dB	–25 dB

Figure 5.8: A two-wire-plus shielded cable is a clear improvement over a single-wire shielded cable because the current return path is decorrelated from the shield. The configuration in D1 is better than the others.

winner (Figure 5.6, configuration B4), you'll notice that the performances are better and especially improved at high frequencies: 42 dB attenuation at 20 MHz. The other configurations (Figure 5.8, configurations D2–D4) aren't bad, but they aren't as good as the configuration D1 in Figure 5.8.

Lastly, I tested an unshielded twisted pair (see Figure 5.9). Once again, the best configuration is when the return wire is grounded only on the receiver side, but the measurement results are really impressive (see Figure 5.9, configuration E1). At 200 Hz, the noise attenuation is even better than with a two-wire untwisted shielded cable (–77 dB versus –72 dB). The attenuation is similar at 2 kHz, but the shielded cable is by far the winner at higher frequencies. In particular, the unshielded twisted

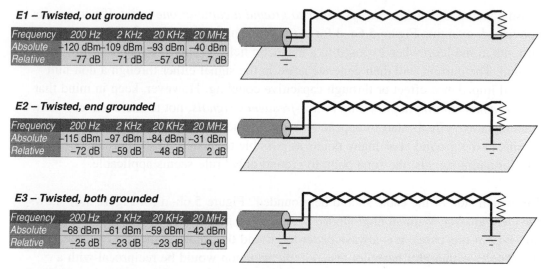

E1 – Twisted, out grounded

Frequency	200 Hz	2 KHz	20 KHz	20 MHz
Absolute	−120 dBm	−109 dBm	−93 dBm	−40 dBm
Relative	−77 dB	−71 dB	−57 dB	−7 dB

E2 – Twisted, end grounded

Frequency	200 Hz	2 KHz	20 KHz	20 MHz
Absolute	−115 dBm	−97 dBm	−84 dBm	−31 dBm
Relative	−72 dB	−59 dB	−48 dB	2 dB

E3 – Twisted, both grounded

Frequency	200 Hz	2 KHz	20 KHz	20 MHz
Absolute	−68 dBm	−61 dBm	−59 dBm	−42 dBm
Relative	−25 dB	−23 dB	−23 dB	−9 dB

Figure 5.9: A simple twisted cable is a good solution at low frequencies. Unfortunately, the lack of a shield is a problem at higher frequencies. That's why shielded twisted cables are a good mix.

pair doesn't seem to provide any significant attenuation at 20 MHz (−7 dB), whereas the two-wire shielded cables provide −42 dB. Unfortunately, I didn't have a shielded twisted cable on hand that day, but I assume that such a cable would add the benefits of a two-wire shielded cable (see Figure 5.8) and the benefits of an unshielded twisted pair (see Figure 5.9). This cable would then be as good as a twisted pair at low frequencies and as good as a shielded cable at higher frequencies—a definitive winner. That's why shielded twisted pairs are everywhere in networks and industrial applications.

Rules to Keep in Mind

After performing these experiments, I concluded that the theory was correct, which is great. The first thing to keep in mind is that *the current return path of any signal should always be kept as close as possible to the signal to reduce inductive-coupling effects.* This can be done either by routing the active wire as close as possible to a ground plane or by disconnecting (if possible) the signal reference from the ground and using a two-wire cable that is ideally shielded and twisted (see Figure 5.7, configuration C2).

I also concluded that *it is always better to ground a cable at one point.* If you are not convinced, examine Figure 5.6c–d for examples. Any "ground loop" will generate currents in the loop when exposed to a magnetic field so noise current will flow in the ground. The current will then generate noise in the signal either through a non-null ground impedance effect or through capacitive coupling. However, keep in mind that *this recommendation applies only to low-frequency signals,* not to radio waves where standing wave effects start to appear. At radio frequencies, it is usually better to connect shields to the ground at as many points as possible to form a meshed ground structure. But for audio signals, the "one point to ground only" rule seems applicable.

But which side of the cable should be grounded? Figure 5.6b–d (or Figure 5.8) shows the clear answer. With an ungrounded source and a grounded receiver (the spectrum analyzer in this case), it is always better to ground the shield at the receiver end, not at the source. Although I have not tested it, the situation would be reciprocal with a grounded source and an ungrounded receiver. The shield should then be grounded at the source side. Figure 5.10 shows these configurations. If you check again, you'll see that this configuration once again reduces the ground loop size.

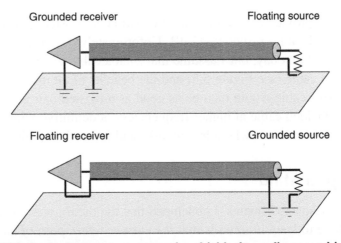

Figure 5.10: This is the best way to ground a shield, depending on which side of the connection is grounded. If both ends are grounded, the performance will be worse, and you'll need to test to see if it is better to ground the shield on one side or both. This may depend on the environment.

All of these tests were done with a victim cable exposed under a fixed high magnetic flux for immunity concerns. Keep in mind that the same analysis is applicable if you need to reduce magnetic flux emitted by a noisy wire (e.g., to comply with spurious emission restrictions). If a given configuration is immune to external fields, it will not generate any field if the internal wire is excited. This is the beauty of reciprocity.

Wrapping Up

I had fun performing the aforementioned experiments. Although some of the results surprised me, they always ended up being compatible with the underlying theory—and that's good news for both the theory and my sanity. The results don't exactly match those published in the original work of Henry Ott, but they are qualitatively close. Once again, this demonstrates that you always need to think twice before putting an old book in your basement. I hope you are now tempted to do your own shielding experiments. I sure was after reading Ott's book. Cable shielding is no longer on the darker side for me, and I hope it isn't for you.

Part 3
Signal Processing

CHAPTER 6

The Fast Fourier Transform from A to Z

Before digging into the topic of digital filters in the next chapters, let's start by refreshing our neurons on the Fourier transform. As you may have noticed, I'm French, so I can't resist telling you that Jean Baptiste Joseph Fourier (1768–1830) was a French mathematician and physicist. His life was quite busy. He was deeply involved in the French Revolution and kept his head on his shoulders only by chance. He then went with Napoleon during his expeditions and became governor of Lower Egypt. Later he returned to Europe and to science and published his well-known paper on thermodynamics in 1822, which brings us back to our subject.

Back to Basics

In his paper, Fourier claimed that any function of a variable can be expanded in a series of sines of multiples of the variable. This transformation was named the Fourier transform. There are books just on the mathematical theory behind the concept, but I will focus on the "engineer's view." For an engineer, a function is usually a table of discrete numbers, so I will never use the real Fourier transform. I will use its discrete time equivalent, the discrete Fourier transform (DFT). I will also restrict my presentation to "real" signals (as opposed to Fourier transforms of complex numbers), and for easier reading I will skip a number of implementation details, which are available on the companion website.

© Elsevier Inc.
DOI: 10.1016/C2009-0-20196-6

Suppose that you worked on some kind of digitizer design and got 1024 samples of a real-life signal stored in RAM with a sampling rate of 250 Ksps. As you probably know, this sampling rate limits your knowledge about the signal to frequencies below the Nyquist limit:

$$F_{NYQUIST} = \frac{F_{SAMPLING}}{2} = \frac{250\,Ksps}{2} = 125\,KHz$$

A DFT takes the 1024 time domain samples and converts them into their frequency domain representation from DC to 125 kHz, with a frequency resolution of

$$\Delta F = \frac{F_{SAMPLING}}{N_{SAMPLES}} = \frac{250\,Ksps}{1024\,points} = 244.14\,Hz$$

You started with 1024 samples and got only 512 frequency bins, that is, 125,000 Hz divided by 244.14 Hz (see Figure 6.1), plus a DC value that is frequency zero. Why half the number of samples? The trick is that the outputs of the DFT are not real numbers. They are complex numbers with two components for each frequency. A complex number is equivalent to a magnitude plus a phase (have a look at Appendix B if you need to know more about complex numbers). So you started with 1024 numbers,

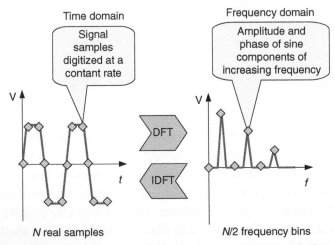

Figure 6.1: A DFT allows you to switch from time domain signals to frequency domain signals through the amplitude and phase of their frequency components.

and you got 2 times 512 numbers, the same quantity of information but from a different viewpoint: frequency domain.

What does it mean? The DFT operation transformed your 1024 time domain digitized samples into a sum of a DC value plus 512 sinusoidal signals with respective frequencies of 244.14 Hz, 488.28 Hz, 732.42 Hz, and so on, up to 125 kHz, each with a given amplitude and a given phase. It is important to understand that this is not an "approximation"; it is just another way of seeing the same signal. The best proof is that the DFT has its reciprocal, the inverse discrete Fourier transform (IDFT), which returns your original signal from its spectral representation. And an IDFT is nothing more than a DFT with a different scaling factor somewhere.

Play with FFT

Why not experiment? I will use Scilab, which was introduced in Chapter 1. Remember that Scilab is an open-source numerical calculation tool. I love it because it works under the operating system of your choice, it is supplied with plenty of compatible toolboxes for signal processing and graphics, and it is free (*http://www.scilab.org*). Have a look at Appendix A for a quick tutorial. Let's write a short Scilab script, which actually decomposes a very short signal into its Fourier coefficients, and reconstructs the signal as a sum of sines:

```
// Just take a 8 points signal
-1->signal=[4.4 2.3 -3.5 -1.2 -2.8 2.0 1.6 2.5]
 signal =
   4.4   2.3 - 3.5 - 1.2 - 2.8   2.   1.6   2.5
// Calculate its Fourier transform, giving 8 complex numbers
-1->spectrum=fft(signal)
 spectrum =

      column 1 to 3
  5.3       10.028427 + 7.5041631i      3.5 - 3.i

      column 4 to 6
  4.3715729 - 2.6958369i   - 5.9      4.3715729 + 2.6958369i

      column 7 to 8
  3.5 + 3.i       10.028427   - 7.5041631i
```

There is a trick here: The output of the FFT contains 8 elements, not 5 as expected (DC plus 4 frequencies). This is because I've tried to hide a mathematical detail from you that I now need to explain: For a mathematician, or for Scilab, the FFT operation in fact takes N complex numbers as input and outputs N complex numbers. But if the N input samples are real numbers, as in our example, then only half of the output numbers are useful; the other half are complex conjugates of the first half. Look at the output just presented.

The first number, 5.3, is the DC component of the signal. It is always a real number. The next four values are the four amplitudes of the frequency components 1, 2, 3, and 4, respectively. They are complex numbers (giving a phase and an amplitude), except the last one, which happens to always be a real number. The next three coefficients are exactly equal to the previous ones, with a negative sign for the imaginary parts. These are often called "negative frequency" components and can be discarded.

We can then calculate the amplitudes and phase of each useful frequency component:

```
// Takes half of the spectrum
-1->usefulspectrum=spectrum(1:$/2+1);

-1->amplitudes=abs(usefulspectrum)/8
 amplitudes  =
    0.6625    1.5656559    0.5762215    0.6419962    0.7375

-1->phases=imag(log(usefulspectrum))
 phases  =
    0.    0.6424053  - 0.7086263  - 0.5525899  - 3.1415927
```

Some explanations: The "$" in Scilab means that the number of elements, that is, "spectrum (1:$/2+1)," gives the first half of the spectrum vector plus one element. The amplitudes need to be divided by the number of input points for a proper scaling, here 8. The trick used to calculate the phase of the complex numbers is based on the fact that the "log" function of a complex number provides the phase as the imaginary part of the result. Just consider that imag(log(x)) is an easy way to get the phase of x.

Now that we have finished the Fourier transform of our initial data set, let's summarize: We started with an arbitrary signal defined by 8 samples (4.4, 2.3, −3.5, −1.2, −2.8, 2.0,

1.6, and 2.5), and we got a DC value (0.6625) and four frequency components with amplitudes of 1.5656, 0.5762, 0.6419, and 0.7375, respectively. These amplitudes each have a given phase: 0.6424, −0.7086, −0.5525, and −3.1415 radians, respectively.

If Fourier was right, our initial signal must be equal to the DC value plus the four sine waves. Do you want to check? Let's create a time step vector, calculate the actual sum of the four sines, and ask Scilab to plot the result and compare it with the original 8 points:

```
// Generate an angle vector from 0 to 2.pi with 80 steps
 x=2*%pi*(0:80)/80;

// Generate the four sines with good amplitudes and phases
f1=2*amplitudes(2)*cos(x+phases(2));
f2=2*amplitudes(3)*cos(2*x+phases(3));
f3=2*amplitudes(4)*cos(3*x+phases(4));
f4=amplitudes(5)*cos(4*x+phases(5));

// Sum them
f=f0+f1+f2+f3+f4;

// Plot the result in a 3 part graph
subplot(1,3,1); plot(signal,'black');
xtitle('A simple 8-points signal');
subplot(1,3,2); plot(f1,'green'); plot(f2,'green');
 plot(f3,'green');
plot(f4,'green'); plot(f,'red');
xtitle('DFT decomposition and sum of sines');

// Compare original signal and recomposition
subplot(1,3,3); plot(f,'red'); plot2d([0:7]*10+1,signal);
xtitle('Original and recomposed signals compared'),
halt ();
```

Look at the rightmost graph in Figure 6.2: There is a perfect match between the original time domain samples and the sum of sines, of course, only at the 8 precise points where the input values were defined. That's the magic of the Fourier transform.

Some Fourier transforms of simple signals are illustrated in Figure 6.3. Take a look at them to get a feeling for a frequency spectrum.

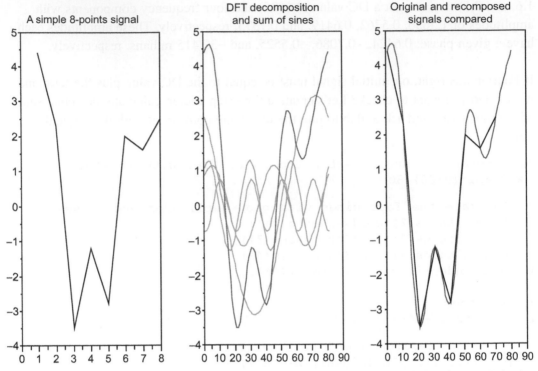

Figure 6.2: This is the output of this example when run in Scilab. The leftmost graph is the original signal, an arbitrary signal defined on 8 points. The middle graph shows the four sinusoidal components calculated by the FFT (*light gray*), with frequencies 1/T, 2/T, 3/T, and 4/T. T is the duration of the signal, as well as its sum (*dark gray*). On the right is the superposition of the original signal and its recomposition.

DFT and FFT

Just a word on the calculation method. The formula used to calculate a DFT is actually quite simple. If the input, time domain, and samples are noted x_n, then the frequency coefficients X_n are calculated as

$$X_k = \sum_{n=0}^{N-1} x_n e^{\frac{-(2\pi i)}{N}kn}$$

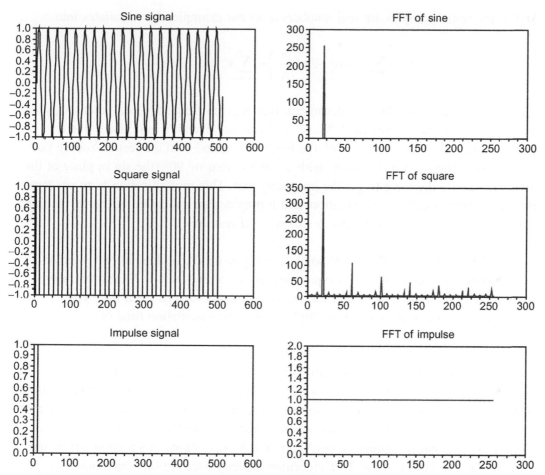

Figure 6.3: The Fourier transform of a sinusoidal signal is just one frequency on the spectrum, where the spectrum of a square signal contains all odd harmonics of the original frequency. A very interesting case: If the input signal is a signal pulse (input signals are all at 0 except one sample at 1), the output spectrum is flat, with all frequencies at the same amplitude.

And, if the initial samples are real numbers as in our example, this translates into

$$X_k = \sum_{n=0}^{N-1} x_n \cos\left(-2\pi k \frac{n}{N}\right) + i\sum_{n=0}^{N-1} x_n \sin\left(-2\pi n \frac{k}{N}\right)$$

Look twice at this formula. In order to calculate a given frequency coefficient X_k, you first need to generate a numerical "wave" at frequency k with a null phase (the cos term), multiply it term per term with the input vector, and sum the result. You've got the real part. Then you do the same with a wave shifted by 90° (the sin in place of the cos), and you do the same to get the imaginary part. This multiplication term per term and summation is called a convolution, and it provides an indication of the correlation factor between the signal and a given sine wave (Figure 6.4).

So, to calculate a full DFT of an N-point vector, you need to calculate two N-point convolutions for each of the $N/2$ frequencies, giving a total of $N \times N$ multiplications. The DFT algorithm then has a calculation complexity of $O(N^2)$. Each time you double the number of points, you multiply by four the calculation time of a DFT.

You have probably seen the abbreviation "FFT," which stands for fast Fourier transform. What is the difference between DFT and FFT? Only the algorithm, not the result. In fact, an FFT is just a common and efficient way of calculating a DFT on a computer or microprocessor. It is not a new concept. Basically, an FFT gives the same result as a DFT, but it requires $N.\log(N)$ and not N^2 operations. It's a great advantage when N starts to increase. The FFT algorithm cleverly uses the fact that some intermediate calculations can be reused for several coefficients because the sine waves are periodic.

I will not detail the FFT algorithm here, but the basic idea is to recursively split the N points into two $N/2$ sets, as there is a mathematical way of recombining the Fourier transform of both halves to obtain the full transform. Doing it again and again, you end up with the transform of two points, which is trivial.

The details of the algorithm are a little harder to explain, but there are plenty of very good implementations available on the Web so you will probably never have to code an

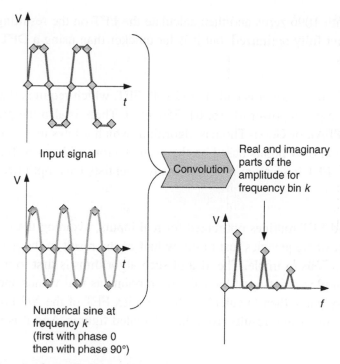

Figure 6.4: Each coefficient of the frequency domain taps is calculated as a convolution between the input vector and a sine of the corresponding frequency with two phases, 0° and 90°, to obtain the real and complex parts of a given frequency coefficient.

FFT yourself. One of the most efficient is the well-known FFTW (see *http://www.fftw. org*), but you will also easily find some source code for your preferred processor, from basic C code versions to hand-optimized assembly code.

FFT Variants

The easiest way to code an FFT is called "radix-2" FFT. It is usable only when the number of points is a power of 2. That's why the vast majority of available FFT codes are optimized for 256, 512, 1024 points, and so forth. Such an FFT algorithm can also be used for non-power-of-2 samples, but the result will be less efficient. For example, one of the common methods of calculating an FFT on a 3000-point vector is simply to pad it to 4096 with 1096 zeros and then calculate the FFT on the resulting 4096 points. This method is not fully optimized, but it is far quicker than using a DFT algorithm on 3000 frequencies.

Another very effective variant is called "radix-4" FFT, which requires that the number of points be 4 to a given power (4, 16, 64, 256, etc.). There is also the prime-factor FFT algorithm (PFA), or Good–Thomas algorithm, which allows users to optimize the FFT calculation of N points based on the prime factors of N, as well as tens of other variants. FFT is so often used that many scientists have optimized it over the years.

You will also find FFT routines optimized for real inputs, avoiding the need to calculate twice the number of frequencies just to throw half of the result in the garbage bin, as we did in our previous example. The idea of such algorithms is first to pack the N real points as $N/2$ complex points, just by using even points as real values and odd points as their imaginary parts, then to calculate the complex FFT of the $N/2$ points, and finally to recover the actual results from the calculated figures. And it is indeed possible.

Windowing

Usually you will need to take a snapshot of the real world, say samples during some tens of milliseconds, and calculate the FFT of these samples to know the frequency components of the signal. If you just take N successive samples, you will in fact apply a rectangular mask on the real-life data, cutting the timeline between the start and the stop of the acquisition. This will work, but the result is that the frequency peaks will be a little distorted just because the input signal will not be infinite.

There is a way to reduce these artifacts: You can, before the FFT, multiply your input data vector with a windowing function that will slowly reduce the amplitude of the signal to zeros on both ends. The result will be less distortion caused by the edges. Several windowing functions exist. The simplest is a linear attenuation of the signal (trapezoidal window), but one of the most used is the so-called Hamming window, which has very good mathematical properties.

Let's try to illustrate windowing with Scilab:

```
N=128;
t = 0:N-1;

// Generates a sine and calculate and plot its FFT
  signal=sin(12.443*2*%pi*t/N);
spectrum=fft(signal);
subplot(2,2,1);
plot2d(signal); xtitle('sine signal');
subplot(2,2,2);
plot(abs(spectrum(1:N/2))); xtitle('FFT of sine');

// Generates a hamming window of the same size N
w=window('hm',N);

// Multiply then element per element with the signal
windowed=signal.*w;

// Calculate and plot the resulting FFT
wspectrum=fft(windowed);
subplot(2,2,3);
plot2d(windowed); xtitle('sine signal, Hamming window');
subplot(2,2,4);
plot(abs(wspectrum(1:N/2))); xtitle('FFT of sine, windowed');
```

The result (Figure 6.5) will show you why windowing must usually be used.

FFT Applications

Why use an FFT? First, of course, to find out the frequency decomposition of a signal, but there are plenty of other useful applications. I explained that a given frequency coefficient is in fact calculated as a convolution between the signal and a reference sine waveform (both cos and sin get the phase as well). Old timers should link this concept to another very useful measurement method: a lock-in amplifier. Such a device allows the measurement of very small sine signals buried in noise through a correlation between the noisy signal and the reference excitation sine signal. This is exactly what a DFT or an FFT does numerically, so such a transform should allow us to get signals out of the noise, and it does.

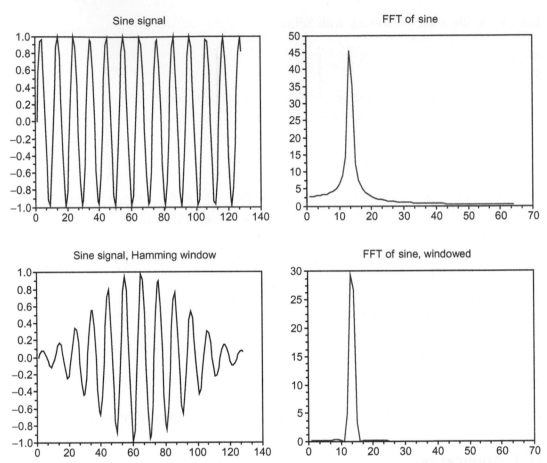

Figure 6.5: The *top* plot illustrates the FFT of a sine signal with rectangular windowing, which means no windowing at all. The same FFT calculated using a Hamming window (*bottom*) shows that the peaks are much more precise.

An example? Suppose that you have a 20-mV sine signal with, unfortunately, 1 V of random noise added to it. A 65,536-point FFT will easily allow you to identify this signal (Figure 6.6).

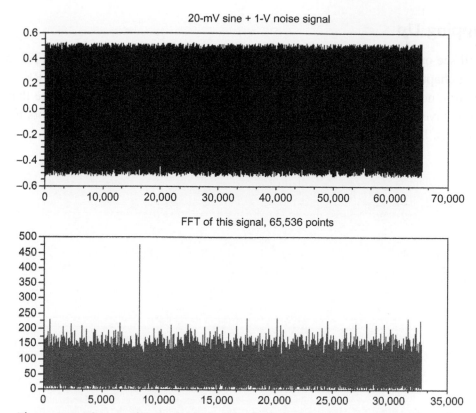

Figure 6.6: The top figure shows a 20-mV sine plus a 1-V random noise. The result seems to be noise. However, an FFT allows this frequency component to be clearly recovered.

```
N=65536;
t = 0:N-1;
signal=20e-3*sin(8212.443*2*%pi*t/N)+(rand(t)-0.5);
spectrum=fft(signal);
subplot(2,1,1);
plot2d(signal); xtitle('20mV sine + 1V noise signal');
subplot(2,1,2);
plot(abs(spectrum(1:N/2))); xtitle('FFT of this signal, 65536
  points');
halt();
xdel();
```

Wrapping Up

We will see other direct or indirect applications of the Fourier transform and FFT in the next chapters, in particular Chapter 13, where I will introduce the OFDM RF modulation now used in many wireless and wireline standards. Anyway, I hope that you now have a better idea of the wonderful tool that Fourier and his successors have provided us. Let's use it in the next chapter!

CHAPTER 7

No Fear with FIR: Put a Finite Impulse Response Filter to Work

Based on the discussion of the Fourier transform in Chapter 6, I can now describe a well-known and very powerful class of digital filters: finite impulse response filters (FIRs). As usual, I will use plain English to explain the ideas, but some mathematical concepts will still appear from time to time. However, I will try to stay as close to the bench as possible, so don't be afraid, breathe normally, and keep reading. If you get lost, just skip to the "FIR for you" section!

Frequency Domain

Imagine that you have an analog signal coming from a real-life sensor. You digitize it by an ADC, with a proper anti-aliasing filter (more on that later), and you try to process it on your preferred microcontroller or DSP to measure some signal trends. But, unfortunately, imagine that the resulting signal is too noisy to be useful. After a quick analysis, you find out that the source of your problem is 50-Hz (or 60-Hz) noise perturbing the sensor. As always, it is too late to modify the hardware, but you could still implement a digital filter on the microcontroller to remove this noise, in that case, through a digital notch filter. But how do you actually code such a filter?

Remember the previous chapter: Thanks to the Fourier transform you know that you can convert a time domain signal into its frequency spectrum. This will give you a first straightforward digital filtering method and will allow me to introduce FIR filters.

Note: This chapter is a corrected and enhanced reprint of the article "The darker side: No fear with FIR: Put a finite impulse response filter to work," *Circuit Cellar*, no. 207, October 2007.

© Elsevier Inc.
DOI: 10.1016/C2009-0-20196-6

Filtering a signal is just modifying its frequency spectrum, enhancing the sub-bands you are interested in or attenuating the others. That's really easy with a discrete Fourier transform (DFT) with a three-step process:

- The first step is to take your input time domain signal and calculate its Fourier transform through a DFT or its speedy FFT variant.

- The second step is to modify the spectrum as you wish, removing, attenuating, or increasing the power of all frequency bands. This can be done easily by multiplying, term by term, the spectrum and a frequency mask. For example, to implement a low-pass filter, you use a frequency mask equal to 1, from DC up to the cutting frequency, and then null.

- You now have a filtered spectrum and you can get back to a filtered time domain signal with the third step, an inverse Fourier transform (IDFT or IFFT).

For those of you who like equations, if S is the signal and M is the wanted frequency response, i.e., the frequency mask, then the filtered signal S' is given by

$$S' = IDFT(DFT(S) \times M)$$

Let's try it in Scilab, on the example of a low-pass filter:

```
// build a signal sampled at 500hz  containing two pure frequencies
//  at 50 and 90 Hz
sample_rate=500;
t = 0:1/sample_rate:0.4;
N=size(t,'*'); //number of samples
signal=sin(2*%pi*50*t)+sin(2*%pi*90*t+%pi/4);
  spectrum=fft(signal);
f=sample_rate*(0:(N/2))/N; //associated frequency vector

// the fft response is symetric we retain only the first N/2 points
n=N/2+1;

// Plot the signal and its FFT
clf()
subplot(3,2,1);
plot2d(f,abs(spectrum(1:n))); xtitle('FFT of original signal');
subplot(3,2,2);
plot(signal); xtitle('Original signal');
```

```
// Build a rectangular frequency mask, low pass, and plot it
lim=26;
filter=[0:N]*0;
filter(1:lim)=1;
subplot(3,2,3); plot(filter(1:N/2)); xtitle('filter in the
  frequency domain');

// Makes it symmetrical (as the FFT is symetrical)
filter(N-lim+2:N+1)=filter(1:lim);

// calculates its fft, which is the impulse response
fftfilter=fft(filter);
impulse=real(fftshift(fftfilter));
subplot(3,2,4); plot2d(impulse); xtitle('Impulse response of
  filter');

// Multiply clements per element the spectrum of the signal and the
  filter mask
filteredspectrum=[1:N];
filteredspectrum=spectrum(1:N).*filter(1:N);

// Plot this filtered spectrum
subplot(3,2,5); plot(abs(filteredspectrum(1:n)));
  xtitle('filtered spectrum');

// Calculates corresponding time domain signal and plot it
filteredsignal=fft(filteredspectrum);
subplot(3,2,6); plot(real(filteredsignal)); xtitle('FFT on the
  filtered spectrum');

halt;
xdel();
```

Start Scilab and load this script. The result is shown in Figure 7.1.

The real strength of the method is that virtually any linear filter can be implemented with the same algorithm. You are not limited to simple filters, such as low-pass filters. Nothing forbids you from designing a more complex filter with several sub-bands and a different behavior in each sub-band. You just have to design a frequency mask accordingly. The calculations will be the same and you will also be able to modify the filter very easily, just by changing the frequency mask.

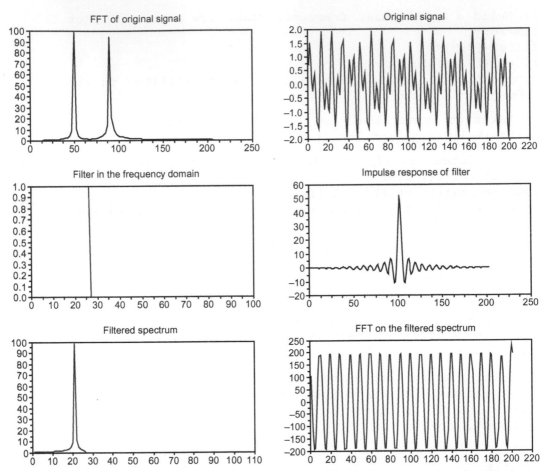

Figure 7.1: A low-pass filter can easily be constructed in the frequency domain. The first line shows a test signal synthesized as the sum of two sines, both in the time (*right*) and in the frequency (*left*) domains. The second line shows a low-pass rectangular frequency mask, as well as its Fourier transform, which is the impulse response of the filter. The last line shows the multiplication of the spectrum, the frequency mask, and its inverse Fourier transform, which is the filtered signal.

From Frequency to Time

This frequency domain filtering method using a DFT and an IDFT is great and effectively used, but it has two problems. First, a DFT (or IDFT) requires a significant amount of processing power and memory. Second, the process is block oriented. You must wait until you have a full buffer of data before calculating the DFT of this buffer. This introduces a long delay in processing, as well as difficulties on the block's boundaries, requiring windowing techniques if you want to filter a continuous signal. Can it be simpler? Yes, but this will, unfortunately, require a very powerful mathematical result. If you have two signals, A and B, then it can be shown that the following is true:

$$IDFT(DFT(A) \times DFT(B)) = A \otimes B$$

What is this crazy $\otimes$? It is what mathematicians call a convolution. It is nothing more than a couple of multiplications and additions, as presented in Figure 7.2.

From the result we can show that the method I have described for frequency domain filters can be done without a DFT or an IDFT! How? Thanks to our frequency domain method, the filtered signal can be calculated as

$$S' = IDFT(DFT(S) \times M)$$

Remember that M is the frequency mask you want to implement. Let's introduce a signal that we will call *IMP*, calculated as the inverse DFT of M:

$$IMP = IDFT(M)$$

So, thanks to the mathematical result above, we have

$$S' = IDFT(DFT(S) \times M) = IDFT(DFT(S) \times DFT(IMP)) = S \otimes IMP$$

Read this formula again: *The filtered signal S′ can, in fact, be directly calculated by a convolution between the original signal S and* IMP, which is simply the IDFT of the frequency mask that you want to implement.

Physically speaking, *IMP* is the signal on the output of the filter when a short impulse is applied on its input. That's why an *IMP* is called the impulse response of the filter. Figure 7.1 shows you the IDFT of the spectrum mask on the middle-left box. That is an

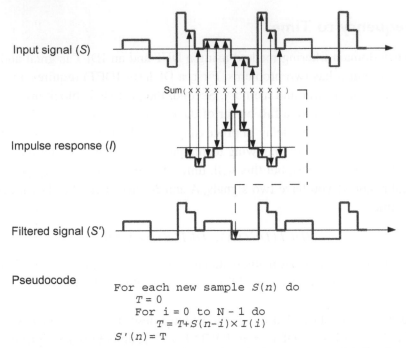

Input signal (S)

Sum (x x x x x x x x x x x)

Impulse response (I)

Filtered signal (S')

Pseudocode

```
For each new sample S(n) do
    T = 0
    For i = 0 to N - 1 do
        T = T+S(n-i)× I(i)
    S'(n) = T
```

Figure 7.2: A convolution between two signals is calculated as N multiplications and one big addition for each output sample. N is the shortest length of the two input signals, usually a fixed time domain mask, which for FIR filters will be the impulse response of the filter. For each output sample, the impulse response is shifted and aligned with the last N samples of the input. The two signals are then multiplied term by term, and all of the results of the multiplications are summed.

IMP. So, with this method, the filtering operation is now based only on time domain signals. You just have to calculate the convolution of the signal to be filtered and the impulse response of the desired filter. Of course, because the frequency mask is a finite-length dataset, the impulse response calculated as its IDFT is also finite. The impulse response is time-limited. That's why the algorithm is called FIR, which means finite impulse response: Contrary to an analog filter, the output of an FIR digital filter damps out and "ignores" any input signal prior to a number of samples equal to the number of samples of the impulse response, which is called the number of taps of the filter.

The main point here is that the *IMP* signal can be precalculated (with an IDFT) when you design your filter. Then, the real-time calculation of the filter doesn't need the DFT at all. If you have been following me so far, you should see that the FIR algorithm gives exactly the same result as the frequency domain digital filtering process we discussed first (Fourier transform, then multiplying the frequency spectrum by a filter mask, then inverse Fourier transform). This is not another method, just a far more effective implementation, especially for time-continuous signals.

FIR for You

You should now have an idea of where a FIR filter comes from. Let's stop focusing on the theory and see what you will actually have to do to implement a FIR digital filter in your next design. It will be an easy, seven-step process (see Figure 7.3).

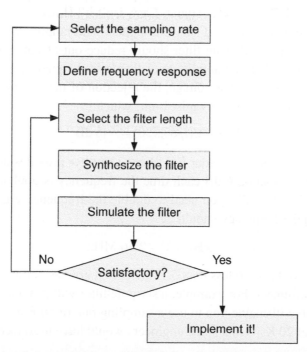

Figure 7.3: The implementation of a FIR filter in your project is a seven-step process. The key is to spend as much time as required in the simulation steps in order to be 99% sure that the filter will do what you want it to do.

Anti-aliasing

The first step is to define the overall sampling frequency, which will usually be the clock frequency of your ADC. This task may sound trivial, but it is often the most important step in the design process. This is not an easy step, so I will spend a little time on it. As you know, thanks to Nyquist, the sampling frequency of your ADC should be at least two times higher than the maximum frequency of the input signal. However, this would require a perfectly sharp anti-aliasing low-pass filter with 100% transfer up to the cutoff frequency, and then 100% attenuation, which doesn't exist anywhere else other than in the minds of mathematicians.

In real life, the definition of the sampling frequency will be a difficult cost balance between an expensive anti-aliasing filter and an expensive, quick ADC and quick digital filter. Still not convinced? Take a numeric example. Suppose that you need a perfect 8-bit digitization of a DC-to-10-kHz signal. Easy, isn't it? However, you want to get 8 bits of precise measurement, so void spurious signals with amplitudes higher than one bit. This means that the anti-aliasing filter should reduce out-of-band signals to levels that are 256 times lower than the amplitude of the desired signal. In terms of power, because power is the square of the voltage, the rejection of the anti-aliasing filter should be better than 1/65,536 above the Nyquist limit, which is

$$10 \times \log(1/65,536) = -48 \text{ dB}$$

Do you plan to use a simple first-order RC filter? The filter provides an attenuation of only 6 dB per octave, meaning 6 dB each time the frequency is doubled above the cutting frequency. To get 48 dB, you need to double the frequency eight times. This means that the Nyquist frequency should be

$$10 \text{ kHz} \times 2^8 = 2.56 \text{ MHz}$$

The sampling rate should be no less than 5.12 Msps! Of course, better filters will enable you to limit the requirement. For example, a six-pole filter will provide an attenuation of 36 dB/octave, which will enable you to use a sampling rate of 50 Ksps, but that is still far above the theoretical 20 Ksps that some engineers would have used (see Figure 7.4).

Even with very good anti-aliasing filters and modest requirements (8 bits), the sampling frequency should be significantly above the double of the input signal frequency, and it

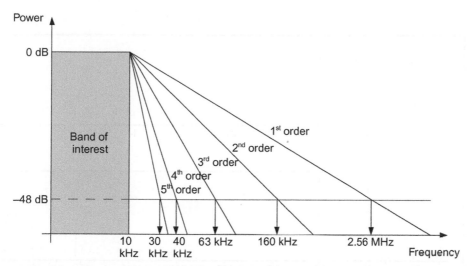

Figure 7.4: This selection of the overall sampling frequency is a cost-optimization process between an ADC and an anti-aliasing filter. For example, to get an 8-bit spurious free-resolution on a 10-kHz signal, the filter rejection above the Nyquist limit should be better than 48 dB, which translates into a Nyquist frequency above 30 kHz even with a good fifth-order anti-aliasing filter giving 60 Ksps.

may be 10 times higher if you want to use simple filters. This is why oversampling is so often a good engineering solution. Think about it for your next project!

Filter Frequency Response

The second step in the FIR design process is far easier and quite fun. It consists of defining the theoretical frequency response that you dream of having. Take a sheet of paper and draw a frequency scale from DC to the Nyquist frequency (half the sampling frequency) and draft the desired filter response. Assume that you need a complex filter such as that in Figure 7.5a. The first box is a filter with two passbands with respective gains of one and two.

The third step involves defining the number of taps of the FIR filter. This is often a trial-and-error process, but you will have a starting point. Remember our discussion about frequency domain filters? Because the FIR taps are calculated from the frequency mask

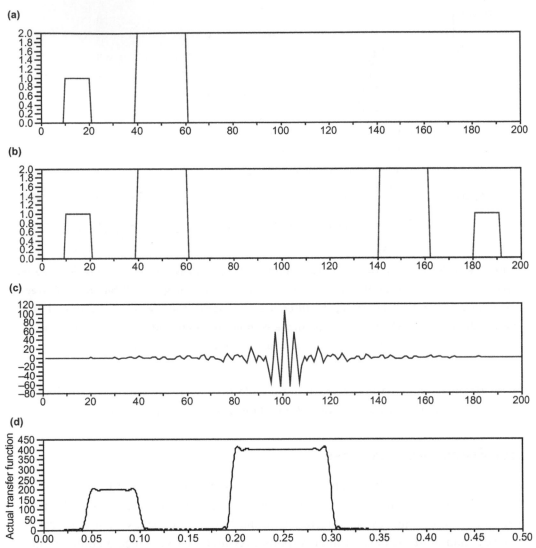

Figure 7.5: This is the output of the Scilab code shown on page 105 and its simulation. From top to bottom, it shows the desired frequency response; the frequency response extended in negative frequencies; its inverse FFT, which is the impulse response of the filter (i.e., the coefficient of the FIR filter); and the actual simulated frequency response.

through an inverse Fourier transform, the number of taps of the filter is equivalent to the number of bins on the frequency mask. So the more complex the target frequency response, the higher the tap count. For example, if you want to design a crude low-pass filter with an "easy" cutoff frequency somewhere in the middle of the passband, an 8-tap FIR filter may be enough. However, if you need a bandpass of 1/100 of the frequency band, you will need significantly more than 100 taps (possibly 500 or more). Figure 7.6 illustrates the frequency response of a simple low-pass filter for different numbers of taps. For a customer who wants a razor a filter, select a filter with 200 taps or more.

Synthesis!

The fourth step is the magical one: the synthesis of the FIR filter. Magical, but understandable after the first part of this chapter. The synthesis of the FIR filter means that you need to calculate the coefficient of each tap. The coefficients are the values of the inverse Fourier transform of the frequency response. *Remember: IMP is the IDFT* of the frequency mask *M*. To calculate the coefficients, you need a tool like Scilab to create a table with as many points as the number of taps to fill its first half with the frequency response from DC to the Nyquist frequency, and to duplicate it upside down in the second half. (Refer to Figure 7.5b.) The duplication is just because the inverse Fourier transform assumes that the input frequencies are both positive and negative as we saw in Chapter 6. Next, you call the inverse Fourier transform function and get the impulse response of the filter, which is the coefficient of each successive tap of the FIR filter. (Refer to Figure 7.5c.)

. . . and Simulation

Mathematically speaking, the filter will then be perfect. The actual response will be the exact specified response at each of the *N* discrete frequencies used to define it—from DC to 10 kHz with a step of 50 Hz (i.e., 10 kHz/200) in my example. However, in real life it is also very important to know how the filter will behave for frequencies *between* these individual frequencies, especially if the tap count is small. This is the fifth step: simulation.

You can do the simulation yourself with test software in your preferred language, but you will need to generate sine signals of slowly increasing frequencies to filter it with

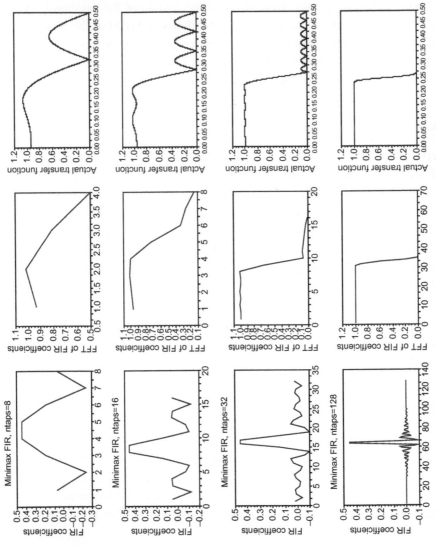

Figure 7.6: The optimized impulse response and the theoretical and actual frequency response of low-pass filters with 8, 16, 32, and 128 taps, respectively. Of course, the filter response improves with higher tap counts, but at the expense of increased computations.

the synthesized FIR and then measure and plot the output amplitude for each frequency. Simple, but painful. The other solution is to use a ready-made frequency-response analysis function available in Scilab: *frmag*.

The entire synthesis and simulation code in Scilab is given below. This is the code that created Figure 7.5. The bottom plot in this figure (d) is the actual response, which is very close to the theoretical response because the number of taps is high.

```
// Specification of the desired frequency response
ntaps=200;
wanted(1:ntaps)=0;
wanted(10:20)=1.0;
wanted(40:60)=2.0;
subplot(4,1,1);
plot(wanted);

// Invert it in the second half (negative frequencies)
for i=1:ntaps/2;wanted(ntaps-i+1)=wanted(i);end;
subplot(4,1,2);
plot(wanted);

// Calculate its FFT, which is the impulse response
fircoeff=real(fftshift(fft(wanted)));
subplot(4,1,3);
plot(fircoeff);

// Calculate the actual frequency response
[hm,fr]=frmag(fircoeff,1000);
subplot(4,1,4);
plot2d(fr',hm');
xtitle('','','Actual transfer function');
```

I must mention that there are more elaborate synthesis techniques that enable you to optimize the filter for a given number of taps and for a given list of mandatory set points on the frequency response graph. These techniques limit the oscillations of the frequency response between the set points using windowing and optimization algorithms. One of them, using the Minimax algorithm, is supported in Scilab through the *eqfir* function. The following was the method used for the examples in Figure 7.6; the code is also very simple to read.

```
ntaps=[8 16 32 128];

for j=1:4;

    // Generation of an optimal filter
    // Two frequency bands are defined, from 0 to 0.23 x nyquist
        and 0.27 to 0.5 x Nyquist
    // Desired responses in these two bands are respectively 1
        and 0, with weights 1 and 0.1
    hn=eqfir(ntaps(j),[0 .23;.27 .5],[1 0],[1 .1]);
    subplot(4,3,1+3*(j-1));
    plot(hn);
    xtitle('Minimax FIR, ntaps='+string(ntaps(j)),'','FIR
      coefficients');

    // Theoritical transfer function
    tf=abs(fft(hn));
    subplot(4,3,2+3*(j-1));
    plot(tf(1:$/2));
    xtitle('','','FFT of FIR coefficients');

    // Actual transfer function
    [hm,fr]=frmag(hn,1000);
    subplot(4,3,3+3*(j-1));
    plot2d(fr',hm');
    xtitle('','','Actual transfer function');
end;
```

Now that you have a simulated response of the FIR filter, it's time to compare it with
the specifications you were looking for. Is it fully satisfactory? If not, you will have to
repeat the entire process of taking different assumptions, which is the sixth step of the
design process. You will probably need to tweak the number of taps or optimize the
frequency response a couple of times.

Time to Solder

The last step is the easiest. Select your preferred digital platform. You can use a
microcontroller for low-speed applications, a DSP for medium-speed projects, or an
FPGA with plenty of hardware multipliers for high-end ones. Code the FIR algorithm,
which should need only a couple of source code lines and some nonvolatile memory
to store the FIR coefficients. To make your life easier, a skeleton of a FIR algorithm

coded in C language is given in the following sidebar and is supplied on the companion website, as well as in a Microsoft Visual C++ project file.

```
//-------------------------------------------------------------------
//
// Skeleton of a FIR filter real-time implementation
//
// Robert Lacoste for Circuit Cellar & Newnes, Nov 2007
// Supplied just as an example and without any warranty…
//
//-------------------------------------------------------------------

// Example compiled on a PC under Visual C++
#include "stdafx.h"
#include <math.h>

// This code uses floating point numbers for an easy reading.
// In actual implementations fixed-point calculations are
// usually implemented for performances purposes.
#define DATATYPE double

// Impulse response of the desired FIR filter
// Here a simple 16-taps low-pass filter
#define NTAPS 16
const DATATYPE IMP[NTAPS] =
{
    0.0546193,   -0.1150983,    0.0538538,    0.0573442,   -0.0667382,
   -0.0987851,    0.1434961,    0.4540236,    0.4540236,    0.1434961,
   -0.0987851,   -0.0667382,    0.0573442,    0.0538538,   -0.1150983,
    0.0546193
};

// Circular buffer, which will hold the last NTAPS values of the
   input signal
// The value of LastNCurrent is a pointer on the last input value in
   the buffer
DATATYPE LastN[NTAPS];
int LastNCurrent=0;

// Input signal sampling
// In a real life application this function will need to be replaced by
```

```
// an actual sampling of the input, through ADC of similar
DATATYPE GetInputSignal(void)
{
    // For test we generate a test signal as a sum of a high
    // frequency signal and a low frequency one
    static int i=0;
    double x;
    x=sin((double)i*3.0)+sin((double)i/10);
    i++;
    return((DATATYPE)x);
}

// Initialization of the FIR buffer (which will hold the last N values
// of the input signal)
void InitFIR(void)
{
    int i;
    for(i=0;i<NTAPS;i++)
            LastN[i]=0;
}

// FIR algorithm : for each call the function calculates the
   convolution
// between the last NTAPS value of the input signal and the NTAPS
   values
// of the impulse response. An index is used in order to avoid any
// memory to memory copy of the LastN buffer
DATATYPE CalculateFIR(DATATYPE in)
{
    DATATYPE s;
    int i,index;

    // Store the new value in the circular buffer
    LastNCurrent=(LastNCurrent+1)%NTAPS;
    LastN[LastNCurrent]=in;

    // Calculate the convolution
    s=0;
    for(i=0;i<NTAPS;i++)
```

```
        {
            index=LastNCurrent-i;
            if (index<0) index+=NTAPS;
            s+=IMP[i]*LastN[index];
        }

        // Return the convoluted value, which is the filter output
        return(s);
}

// Main program : just a test…
int main(int argc, char* argv[])
{
    int i;
    DATATYPE in,out;

    InitFIR();
    printf("FIR simulation\n\n");
    for(i=0;i<100;i++)
    {
        in=GetInputSignal();
        out=CalculateFIR(in);
        printf("In= %+5.5f Out= %+5.5f\n",in,out);
    }
}
```

Your FIR filter should work well as soon as you have corrected the usual bugs (at least if you haven't forgotten a good anti-aliasing filter on your PCB).

To answer a question that I often got from customers, it is indeed possible to perform signal processing on a microcontroller, and FIR in particular. A reasonably complex 32-tap FIR filter needs 32 multiplications and 32 additions per sampling cycle, usually done on 16-bit integer numbers.

If you are clever, you will notice that the impulse response is always symmetrical, which enables you to divide the number of multiplications by two. Anyway, even if you select an 8-bit 10-MIPS microcontroller, make sure you use one with an on-chip 8 × 8-bit hardware multiplier (e.g., a PIC18F). You should then be able to implement such a FIR filter with sampling rates up to 40 Ksps or more, and a reasonable number of taps on so small a computing platform.

Wrapping Up

So, digital filters are not only for high-end applications. I hope that I have convinced you that FIR filters are not black magic, even if they are sometimes on the darker side. Try them in your next project!

CHAPTER 8

Multirate Techniques and CIC Filters

In Chapter 7, I talked about finite impulse response (FIR) filters. In a nutshell, FIR filters allow a digital filter to be implemented with virtually any response curve in a DSP, FPGA, or even a microcontroller, but at the expense of a quite significant number of arithmetic operations. For example, to implement a 32-tap FIR filter, you need to perform 32 multiplications and 31 additions per sample, which translates into a lot of MIPS if your application requires a high sampling rate.

After this article was first published in *Circuit Cellar*, a reader contacted me. He wanted to use a high-end DSP to filter a signal sampled at 16 Msps, and he had trouble implementing the required FIR filter. Basically, his input signal was nearly a 5-MHz sine wave, and he needed to analyze the small perturbations of this signal at around 5 MHz, say from 4.95 to 5.05 MHz. I'm sure you know of plenty of situations like this one, for applications ranging from digital radio receivers to ultrasonic systems. Because the signal was 5 MHz, the choice of a 16-Msps sampling frequency made sense.

As you know, to avoid any nasty aliasing, the minimal sampling frequency should be at least twice the highest frequency present in the signal, so in this case 2×5 MHz = 10 Msps. Using 16 Msps allowed my reader to implement a reasonable anti-aliasing low-pass filter, with a low attenuation at 5 MHz and a high attenuation at 20/2 = 10 MHz. Then he needed to implement a digital filter to isolate the frequencies of interest from background noise, meaning a bandpass filter centered at 5 MHz with an

Note: This chapter is an enhanced reprint of the article "The darker side: Multirate techniques and CIC filters," *Circuit Cellar*, no. 231, October 2009.

© Elsevier Inc.
DOI: 10.1016/C2009-0-20196-6

approximately 100-kHz bandpass. Directly using a FIR filter would require at least a 256-tap filter to obtain this resolution as discussed in Chapter 7, and this translates into 256×2 operations executed 16 million times per second, resulting in an 8192-MIPS requirement. It was not surprising that this reader had some difficulty with his DSP.

What does this mean? Simply that the direct use of a FIR filter isn't a very good idea in this particular case! In this chapter I will show you how to deal with such problems, through multirate signal processing techniques and their companion filters of choice: *cascaded integrator comb* (CIC) filters.

Multirate?

Real-life signals are full of information. However, only a very small part of this information is typically useful, and life is far easier if we succeed in reducing this flow of information down to what we actually need, in order to process it efficiently. This is exactly the job done by a good assistant who reads you the daily newspapers and drops a press review on your desk. By the way, you should have a book on information theory on your shelf. (Either the initial publications by Claude E. Shannon, or any of the good vulgarization books on that subject. There are some really fundamental notions in them that would be a very good subject for another chapter.)

Let's go back to our example. The 16-Msps sampling rate was dictated by the 5-MHz frequency of the signal, but the actually interesting information was only included in a 100-kHz-wide channel, so somewhere this application is throwing away 98% of the information. For example, if the target was to simply send the decoded signal to a DAC, then a 200-Ksps sampling rate on the output side would be enough to transmit a 100-kHz-wide signal. So it is fundamentally useless to calculate a filtered value 16 million times per second, as only 2% of the calculated information will be used on the output. How to optimize this situation? Through multirate of course. Note that another approach could be used to achieve roughly the same results: undersampling. However, this would bring us too far from our subject this time.

The initial signal has a high bit rate but a low proportion of useful information. Multirate processing is an algorithmic technique in which you reduce step by step the sampling rate of a signal while keeping the required information. This means it

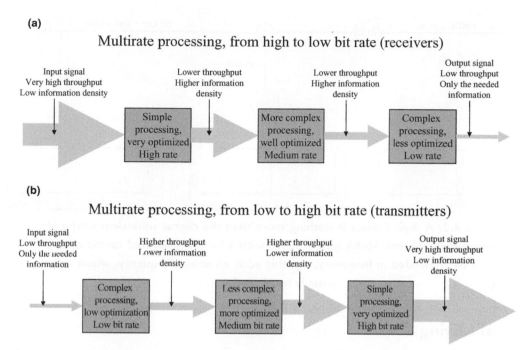

Figure 8.1: (a) Multirate processing is a technique that allows a high sampling rate signal to move gradually down to more manageable speeds while the density of useful information increases. (b) Reciprocally, the same approach can be used in the other direction, usually to build transmitters.

increases step by step the density of information (Figure 8.1a). The trick is to move the algorithmic complexity as far down as possible to lower sampling rates, thus reducing the overall processing requirement.

In a nutshell, the idea is to start with a high bit rate signal, apply algorithms that are as simple as possible, reduce the bit rate, then apply slightly more complex algorithms, and so on, using as many steps as required. The same techniques could be used to increase step by step the data rate of a signal with a reduction in the density of information (Figure 8.1b).

My reader's example is definitely a good candidate for multirate processing. Let's use it as our driving example.

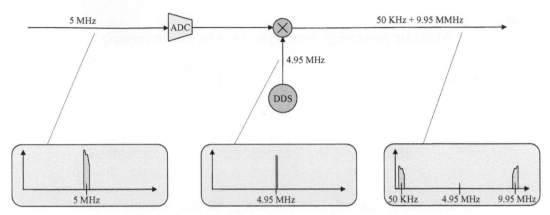

Figure 8.2: A digital mixer is nothing more than the digital equivalent of the old heterodyne system. Multiplying a signal with a locally generated carrier allows it to be translated in frequency, but this adds an image frequency, which must usually be filtered out.

Digital Mixing

Let's start with our 5-MHz signal sampled at 16 Msps. We need to reduce the signal in the processor before executing the computing-intensive analysis algorithms. Usually, the first step is to bring the portion of interest down to a lower frequency but still using the same 16-Msps sampling rate. In an analog radio receiver, such a frequency translation is done using a heterodyne mixer, a technique experimented with by Reginald Fessenden as early as 1900. In the analog heterodyne stage, the input signal is multiplied by a sine local oscillator. A digital mixer is nothing more than the computerized implementation of the same principle (Figure 8.2).

Let's see how such a mixer works. Do you remember your trigonometric formulae?

$$\cos(a) \times \cos b = \frac{1}{2}\cos(a-b) + \frac{1}{2}\cos(a+b)$$

So, if you multiply each sample of a 5-MHz signal with, for example, a 4.95-MHz synthesized local oscillator, you will get a difference signal, which is 5 MHz – 4.95 MHz = 50 kHz (the intermediate frequency, or IF), and a sum signal at 5 MHz + 4.95 MHz = 9.95 MHz (the image frequency):

$$\cos(2\pi5{,}000{,}000t) \times \cos(2\pi4{,}950{,}000t) = \frac{1}{2}\cos(2\pi50{,}000t) + \frac{1}{2}\cos(2\pi9{,}950{,}000t)$$

Note that as this relationship is linear, the spectrum of the signal is maintained. For example, if the original signal included frequency components at 5 and 5.001 MHz, then the down-converted signal will have components at 50 and 51 kHz, respectively.

In practice, such a 4.95-MHz "oscillator" can be implemented as a software-based direct digital synthesis oscillator (DDS), which will require just one addition and one sine table look-up per sample. (I will present DDS in Chapter 11.) You will then have to calculate only one multiplication at each signal sample to do the mixing, so this process will only use a small number of CPU cycles.

Here I must highlight a difficulty: This simple digital mixing scheme works only if you are sure that your input signal has frequency components that are very close to 5 MHz; that is, only from 4.95 to 5.05 MHz. Imagine, for example, that you also have a quite strong input signal of 4.90 MHz; the translated output will then have a component at $4.90 - 4.95 = 50$ kHz. But as $\cos(-x) = \cos(x)$, it will be added to the translated 5-MHz input signal, resulting in a mess. In such a case, you need to filter the input signal before the ADC, increase the intermediate frequency, or use a more sophisticated mixing technique such as IQ mixing. More on that in Chapter 13, but for the moment I will consider that simple mixing is enough for simplicity.

Decimation

Let's summarize where we are. We started with a 5-MHz+/–50-kHz signal and sampled it at 16 Msps with a fast ADC. We then multiplied it with a signal generated by a digital 4.95-MHz local oscillator. This translated the 5-MHz input signal down to a signal centered around 50 kHz (more exactly from 0 to 100 kHz), but we still have two issues: First, we have an image signal at 9.95 MHz, which is inevitably added to our signal of interest; second, the sampling rate is still 16 Msps.

Reducing the sampling rate is known as decimation, and this is really a simple process. Do you want to reduce the sampling rate from 16 Msps down to 200 Ksps? As $16{,}000/200 = 80$, just keep one sample per 80 samples and throw away the other 79!

This is fully legitimate but only if you are sure that the signal can be reliably represented with a 200-Ksps sampling rate, which means that it shouldn't have any frequency component above 100 kHz. And this is why we have a problem with the 9.95-MHz image signal. If we just decimate the mixed signal, the output will be useless as both images will be merged together. We first need to filter out the image, which means implementing a low-pass digital filter.

Are we back to the original problem of implementing a digital filter at 16 Msps? No. As the two signals are now very far apart in frequencies, a quite crude digital filter can be used, which will be more realistic on the computational side.

Moving Averages and CIC Filters

You have probably observed that averaging allows "noise" to be removed in a data set. In other words, an average, and more specifically a moving average (also called a running average), is a crude and simple way to implement a low-pass filter: Just take N consecutive samples of the signal, calculate their average (or their sum, which is the same number multiplied by N), and use the result as the output of the filter. Figure 8.3a shows you an example with $N = 4$.

As a moving average doesn't require anything other than additions, we can use such a filter for our problem: With a proper choice for N, a moving average filter will allow us to filter out the 9.95-MHz image and keep only the 50-kHz signal. Then a decimation step will be legitimate. Of course, it would be useless to calculate all moving averages before decimation. For example, if we want to decimate by a factor of 2, we just need to calculate one average every two samples (Figure 8.3b).

However, look again at Figure 8.3b: This algorithm is not very efficient, as we calculate the same operation several times. For example, the first output value is $x_0 + x_1 + x_2 + x_3$, and the second output value is $x_2 + x_3 + x_4 + x_5$: We have calculated the addition $x_2 + x_3$ two times. The situation will be worse if the average was calculated on a larger number of points, which is usually the case. For example, if we needed to use an average of $N = 256$ points with a decimation factor of $D = 8$, we would calculate each addition 32 times. There should be a more efficient way.

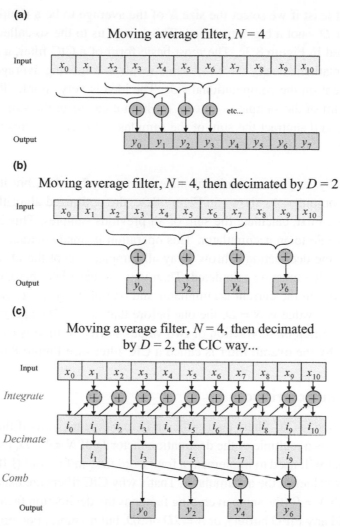

Figure 8.3: A moving average filter (a) is not very effective in terms of calculation even if coupled with decimation and even if only the required averages are calculated (b). For example, here the sum $x_2 + x_3$ is calculated twice. The CIC algorithm (c) is simply a more optimized algorithm to calculate exactly the same values.

Of course it is, at least if we select the size N of the average to be a multiple of the decimation factor D—not a big concession. This brings us to the so-called CIC filter, which is illustrated in Figure 8.3c. The most basic form of a CIC filter, a one-stage CIC ("CIC1"), gives mathematically the exact same result as a moving average filter, but it is far more efficient on the computational side. The idea is very simple. Rather than calculating the sum of the samples x_{101} to x_{110}, why not calculate the sum of the samples x_0 to x_{110} and subtract the sum of the samples x_0 to x_{100}? The result would be exactly the same.

This may seem silly, as you wouldn't calculate it that way by hand, but it is a very good idea for a computer. As shown in Figure 8.3c, the optimized algorithm is a three-step process. First, calculate the sum of all previous samples. This is easy: Just add each new sample to an accumulator. This operation is, mathematically, integration. Then you can do the decimation: Throw away all samples except the D samples for which an output value must be calculated. Then you can calculate this output value as the difference between the current accumulator and one of its previous values; more exactly the previous value is $N = D$, the one before that is $N = 2D$, and so on. In signal processing, the addition or subtraction of two values shifted in time is called a comb, so now you know why the overall filter is called a CIC filter (see Figure 8.6a).

CIC Filter Characteristics

As already discussed, a CIC filter can be efficiently implemented only if the size N of the averaging window is a multiple of the decimation factor D. If $N = 4 \times D$, then the comb section of the filter will need to implement a four-word first-in/first-out (FIFO) buffer to store the last four values of the accumulator. That's why CIC filters are usually implemented with $N = D$ (the same averaging factor as the decimation factor, which allows us to avoid any FIFO buffer), or $N = 2D$ or $3D$, but no more. For simplicity, I will focus only on the simpler case, $N = D$, in the remainder of this chapter.

What is the frequency domain characteristic of such a one-stage CIC filter, meaning its attenuation at a given frequency? It is the same as the characteristic of a moving average filter since they are just two implementations of the same filter. What is that characteristic? You have three ways to find it: intuitively, by simulation, or mathematically. Let's start with the math.

Do you remember Chapter 7 on FIR filters? A moving average filter is in fact nothing more than a special FIR filter with all coefficients set to 1. So you will remember that its frequency response is the Fourier transform of a rectangular window, which is a very common function in signal processing, named sinc(x) and equal to sin(x)/x. Do you want to see its shape? I've written a small Scilab script to plot the frequency response of a $D = N = 8$ CIC filter:

```
// Simulation length
LENGTH = 1024;

// Width of the window (in samples)
D = 8;

// generate the filter impulse response (rectangular, length D)
imp = zeros(1:LENGTH);
imp(1:D) = 1/D;

// calculate its frequency response through FFT
Freqresponse = abs(fft(imp));

// plot it, in linear and log scale !
subplot(1,2,1);
plot2d((1:LENGTH/2)/LENGTH,freqresponse(1:$/2));
xtitle('Moving average filter - D = 8 (linear scale)');
subplot(1,2,2);
db = 20*log10(freqresponse+1e-200);
plot2d((1:LENGTH/2)/LENGTH,db(1:$/2),rect = [0,-25,0.5,0]);
xtitle('Moving average filter - D = 8 (dB)');
```

Download Scilab on your PC if it is not already there. Run this script and you will get the plot shown in Figure 8.4.

How to analyze Figure 8.4 intuitively? There is a first lobe with a low-pass shape as planned. The gain becomes lower and lower as the frequency increases, to a point where the gain is null. For $D = 8$ this point is at a frequency $F = 1/8 = 0.125$ times the sampling rate. Think twice. This is normal: If the input signal has such a frequency, then one of its periods will be exactly as long as the averaging window, so the average will be null. However, when the input frequency continues to increase, the filter is not as good; there is a second lobe with a maximum attenuation of −13 dB, then a new null, then a smaller lobe, and so on. This is the characteristic of a moving average filter:

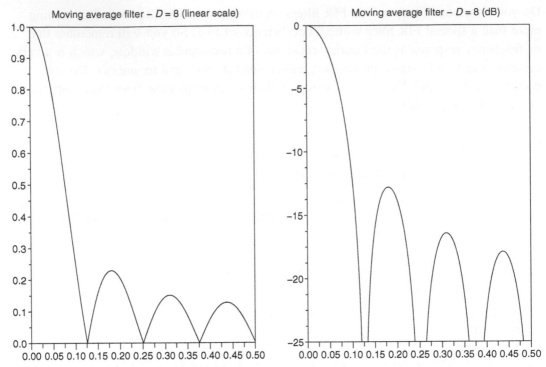

Figure 8.4: Here is the frequency response of a CIC1 filter with an averaging window of eight samples and a decimation factor of 8, both with linear scale (*left*) and logarithmic scale (*right*). This is mathematically speaking a sin(x)/x shape, with a main low-pass lobe but also with significant residual side lobes at higher frequencies.

Each time the period of the input signal is a submultiple of the averaging duration, the attenuation will be maximal.

Such a filter may be enough for your application, but what can you do if the attenuation is not as steep as you want? The first solution would be to increase the size of the averaging, either by increasing the decimation factor D or by implementing a FIFO with $N = 2D$ or $3D$, but this is not always possible because it would alter the bandwidth. The other solution is to cascade several identical filters: Their attenuation will be added, thus significantly decreasing the importance of the side lobes. Let's simulate it in Scilab:

```
//---------------------
// Frequency response of a multistage CIC decimation filter
//---------------------

// Simulation length
LENGTH = 1024;

// comb delay (equal to decimation ratio)
D = 8;

// Input pulse not at start position to simplify coding)
in = zeros(1:LENGTH);
in(100) = 1;

//*********************************************
// Single stage CIC decimation filter
//*********************************************

// Integrator 1
s1 = zeros(1:LENGTH);
for I = 2:LENGTH
  s1(i) = s1(i-1)+in(i);
end;

// Comb 1
c1 = zeros(1:LENGTH);
for i = 1+D:LENGTH
  c1(i) = s1(i)-s1(i-D);
end;

// calculate its frequency response through FFT
Freqresponse = abs(fft(c1))/D;

// And plot it
subplot(3,2,1);
plot2d((1:LENGTH/2)/LENGTH,freqresponse(1:$/2),rect =
  [0,10e-5,0.5,1]);
xtitle(msprintf('CIC filter - D = %d - 1 stage (linear
  scale)',D));
subplot(3,2,2);
db = 20*log10(freqresponse+1e-200);
plot2d((1:LENGTH/2)/LENGTH,db(1:$/2),rect = [0,-80,0.5,0]);
xtitle(msprintf('CIC filter - D = %d - 1 stage (dB)',D));
```

```
//**********************************************
// 2-stages CIC decimation filter
//**********************************************

// Integrator 1
s1 = zeros(1:LENGTH);
for I = 2:LENGTH
  s1(i) = s1(i-1)+in(i);
end;
// Integrator 2
s2 = zeros(1:LENGTH);
for i = 2:LENGTH
  s2(i) = s2(i-1)+s1(i);
end;
// Comb 1
c1 = zeros(1:LENGTH);
for i = 1+D:LENGTH
  c1(i) = s2(i)-s2(i-D);
end;
// Comb 2
c2 = zeros(1:LENGTH);
for I = 1+D:LENGTH
  c2(i) = c1(i)-c1(i-D);
end;
// calculate its frequency response through FFT
Freqresponse = abs(fft(c2))/(D*D);

// And plot it
subplot(3,2,3);
plot2d((1:LENGTH/2)/LENGTH,freqresponse(1:$/2),rect =
  [0,10e-5,0.5,1]);
xtitle(msprintf('CIC filter - D = %d - 2 stages (linear scale)',
  D));
subplot(3,2,4);
db = 20*log10(freqresponse+1e-200);
plot2d((1:LENGTH/2)/LENGTH,db(1:$/2),rect=[0,-80,0.5,0]);
xtitle(msprintf('CIC filter - D = %d - 2 stages (dB)',D));

//**********************************************
// 3-stages CIC decimation filter
//**********************************************
```

```
// Integrator 1
s1 = zeros(1:LENGTH);
for i = 2:LENGTH
  s1(i) = s1(i-1)+in(i);
end;

// Integrator 2
s2 = zeros(1:LENGTH);
for i = 2:LENGTH
  s2(i) = s2(i-1)+s1(i);
end;

// Integrator 3
s3 = zeros(1:LENGTH);
for i = 2:LENGTH
  s3(i) = s3(i-1)+s2(i);
end;

// Comb 1
c1 = zeros(1:LENGTH);
for i = 1+D:LENGTH
  c1(i) = s3(i)-s3(i-D);
end;

// Comb 2
c2 = zeros(1:LENGTH);
for i = 1+D:LENGTH
  c2(i) = c1(i)-c1(i-D);
end;

// Comb 3
c3 = zeros(1:LENGTH);
for i = 1+D:LENGTH
  c3(i) =c 2(i)-c2(i-D);
end;

// calculate its frequency response through FFT
freqresponse = abs(fft(c3))/(D*D*D);

// And plot it
subplot(3,2,5);
plot2d((1:LENGTH/2)/LENGTH,freqresponse(1:$/2),rect =
  [0,10e-5,0.5,1]);
```

```
xtitle(msprintf('CIC filter - D = %d - 3 stages (linear
  scale)',D));
subplot(3,2,6);
db = 20*log10(freqresponse+1e-200);
plot2d((1:LENGTH/2)/LENGTH,db(1:$/2),rect = [0,-80,0.5,0]);
xtitle(msprintf('CIC filter - D = %d - 3 stages (dB)',D));
```

The resulting graph is provided in Figure 8.5.

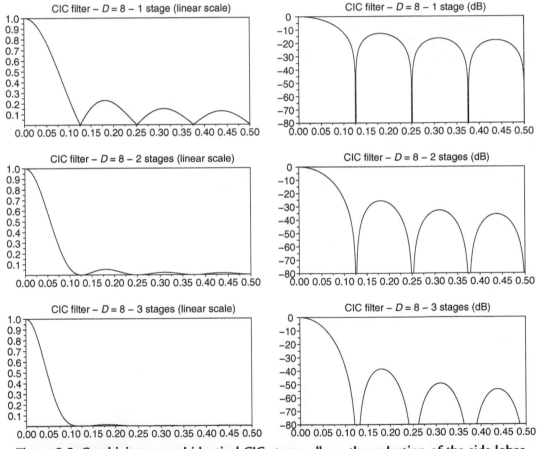

Figure 8.5: Combining several identical CIC stages allows the reduction of the side lobes, at a slightly higher processing cost. Here you have a comparison of CIC1 (*top*), CIC2 (*middle*), and CIC3 (*bottom*).

In practice, there is an efficient way to cascade several CIC filter stages. For example, let's say that you want to cascade two CIC filters with the same parameters N and D just to double the attenuation in the stop band. You will perform the following steps: integrate, comb, then integrate, decimate, and comb. As all these operations are linear, you will get the same result even if you shuffle them around, and a very efficient order is the following: integrate, integrate, decimate, comb, comb. This is the usual implementation of two-stage CIC2 filters, as shown in Figure 8.6b. Of course, nothing prevents you from building CIC3, CIC4, or more complex filters. Just take care of the number of bits of the accumulators, as the numbers can be big.

Two last words before going to implementation: CIC filters can also be used in the other direction, to increase a sampling rate. You just have to reverse the operations: Comb first, then interpolate, then integrate (Figure 8.6c). An interpolator in that context is nothing more than a latch, copying the same value several times on the output. You can check it. With such a CIC filter implemented you will get a linear interpolation between each input sample.

Lastly, a caution: The passband of a CIC filter is all but flat. The attenuation is 0 dB only at DC and then increases up to 3 dB regularly, not abruptly as you would expect with a good low-pass filter. This is the price you pay for efficiency. However, nothing prevents you from correcting this behavior by computation: A usual implementation is to add, after decimation, a FIR filter with a response curve precisely calculated to compensate for the behavior of the CIC filter in its passband, usually a filter named *invsinc* (inverse $\sin(x)/x$). The good news is that such a correcting FIR filter will be on the low bit rate side, which is therefore far easier to implement.

Firmware Implementation

I know that you prefer actual implementation to theoretical presentations, right? In high-speed systems such as software-defined radios, CIC and multirate techniques are usually implemented either in high-end FPGAs or in DSPs. There are also some generic-purpose digital receiver chips such as the AD6640 (Analog Devices), which implements exactly what I have discussed: digital oscillator and mixer, high-speed 65-Msps CIC filter, and correcting FIR filter. The only difference is that the AD6640

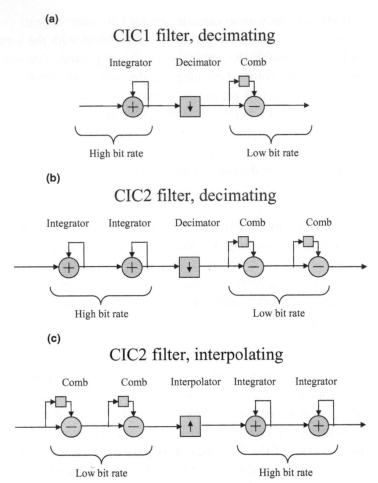

Figure 8.6: (a) A CIC1 filter is nothing more than an integrator, a decimator, and a comb, which is a simple subtraction. **(b)** A CIC2 filter can be built efficiently by grouping the two integrators and the two combs together. **(c)** Reciprocally, a CIC filter can be used to increase the bit rate, just by exchanging the integrator and comb sections and replacing the decimator by an interpolator.

includes a complex mixer and two filter chains in order to be used for digital modulation systems.

However, CIC and multirate techniques can also make your life easier in many low-cost microcontroller-based designs. I have built a basic example for you. Suppose that you need to design some kind of Doppler sound velocity measurement system.

Let's assume that you have a loudspeaker that sends a 7-kHz audio signal in the air and close to it is a microphone to grab the sound bounced back by small moving objects. You want to know the relative speed of the moving objects. You can find out by analyzing the Doppler shift of the received sound: Any object coming closer to the microphone will generate a frequency slightly higher than 7 kHz, and reciprocally any object moving away will generate a sound frequency below 7 kHz.

The frequency shift is simply 7 kHz times the ratio between the speed of the object and the speed of sound in the air, which is about 340 m/s. For example, if the speed of the object is 1 m/s, the shift will be 7 kHz × 1/340 = 20.5 Hz. So if you analyze the frequency spectrum of the microphone signal from 7 kHz − 50 Hz = 6950 Hz to 7 kHz + 50 kHz = 7050 Hz, you will detect and measure the speed of any object in a 2-m/s speed window backward or forward.

A direct frequency analysis of the signal will give a very poor resolution; you need to "zoom" in on this frequency window before performing a detailed analysis with an FFT or similar techniques. It seems to be a good application for multirate processing, doesn't it? Of course, this is exactly the same problem as the one I talked about in the first part of this chapter; I've just moved the operating frequency from 5 MHz down to 7 kHz in order to use a low-speed processor and to illustrate the concept on a simple design.

Figure 8.7 shows you the corresponding design. The input signal (6950 to 7050 Hz) is first digitized at 25 Ksps with an ADC and mixed with a 7.050-kHz digital synthesized oscillator. This translates the signal of interest into low frequencies (0 to 100 Hz), but unfortunately with an image frequency close to 14 kHz.

Then, a CIC2 filters out the image frequency and reduces the sampling rate by a factor of 16, down to 1.562 kHz, which is large enough for a 100-Hz bandwidth. With such a

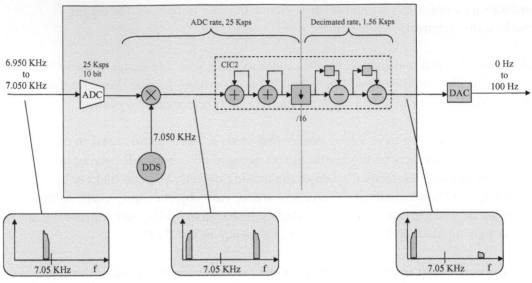

Figure 8.7: This is the architecture of the example I have implemented on a small PIC24 microcontroller. From left to right you recognize a digital mixer and a CIC2 low-pass filter. This design makes it possible to digitize a 7-kHz signal and to zoom out its frequencies from 6950 to 7050 Hz.

frequency range, a low-cost 16-bit processor, such as a PIC24 from Microchip, is enough. Figure 8.8 shows you the full corresponding schematic, which is more than simple.

I coded the corresponding firmware in C. I think that reading it will provide you with more information on CIC filters than a thousand words, so here it is:

```
// CIC filter example
// Target : PIC24FJ64GA002
// (c) Robert Lacoste 2009 - Provided as an example only

#define USE_AND_OR /* To enable AND_OR mask setting */

#include <timer.h>
#include "p24FJ64GA002.h"
#include "stddef.h"
```

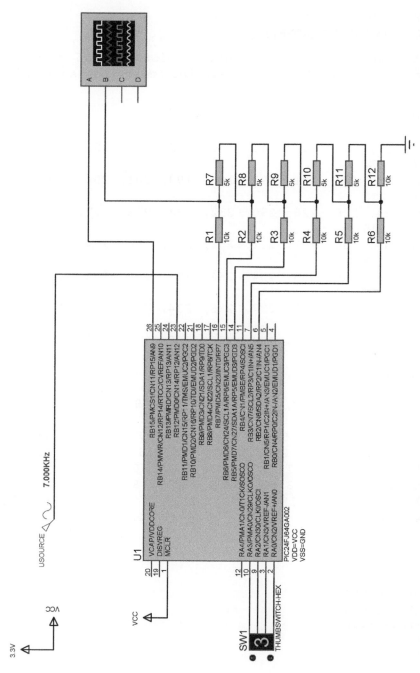

Figure 8.8: The corresponding schematic is straightforward. The input signal is sent directly to an ADC input of a PIC24FJ64GA002. A BCD switch connector to RA0.3 is read by the firmware to select which data to send out on RB0...7: ADC value, DDS value, mixed value, or filtered output. Then a crude 6-bit R2R DAC converts it to analog form. Lastly, a virtual scope displays it. Pin RB15 is just used to signal when the interrupt routine is active.

```
_CONFIG1(FWDTEN_OFF);
_CONFIG2(FNOSC_PRI&POSCMOD_NONE);

unsigned short phase;              // Phase register is 16 bits
unsigned short phaseincrement;     // Phase increment is also 16 bits

// Sine table, from 0 to 255
const unsigned char SINE_TABLE[]=
{
128, 131, 134, 137, 140, 143, 146, 149, 152, 156, 159, 162,
  165, 168, 171, 174,
176, 179, 182, 185, 188, 191, 193, 196, 199, 201, 204, 206,
  209, 211, 213, 216,
218, 220, 222, 224, 226, 228, 230, 232, 234, 236, 237, 239,
  240, 242, 243, 245,
246, 247, 248, 249, 250, 251, 252, 252, 253, 254, 254, 255,
  255, 255, 255, 255,
255, 255, 255, 255, 255, 255, 254, 254, 253, 252, 252, 251,
  250, 249, 248, 247,
246, 245, 243, 242, 240, 239, 237, 236, 234, 232, 230, 228,
  226, 224, 222, 220,
218, 216, 213, 211, 209, 206, 204, 201, 199, 196, 193, 191,
  188, 185, 182, 179,
176, 174, 171, 168, 165, 162, 159, 156, 152, 149, 146, 143,
  140, 137, 134, 131,
127, 124, 121, 118, 115, 112, 109, 106, 103, 99, 96, 93, 90,
  87, 84, 81,
79, 76, 73, 70, 67, 64, 62, 59, 56, 54, 51, 49, 46, 44, 42, 39,
37, 35, 33, 31, 29, 27, 25, 23, 21, 19, 18, 16, 15, 13, 12, 10,
9, 8, 7, 6, 5, 4, 3, 3, 2, 1, 1, 0, 0, 0, 0, 0,
0, 0, 0, 0, 0, 0, 1, 1, 2, 3, 3, 4, 5, 6, 7, 8,
9, 10, 12, 13, 15, 16, 18, 19, 21, 23, 25, 27, 29, 31, 33, 35,
37, 39, 42, 44, 46, 49, 51, 54, 56, 59, 62, 64, 67, 70, 73, 76,
79, 81, 84, 87, 90, 93, 96, 99, 103, 106, 109, 112, 115, 118,
121, 124
};

// Integration registers
signed short integrate1=0;
signed short integrate2=0;
```

```
// Comb registers
signed short comb1=0;
signed short comb2=0;
signed short lastint2=0;
signed short lastcomb1=0;
unsigned char CombCpt=0;
#define COMBRATIO 16

// Timer interrupt routine
void __attribute__ ((interrupt,no_auto_psv)) _T2Interrupt (void)
{
        int ADCValue;
    unsigned char lookupindex;
    unsigned char DDSValue;
    signed char MixedValue;
    signed char FilteredValue;

    // Clear interrupt flag
       T2_Clear_Intr_Status_Bit;

    // Signal interrupt processing on RB15 pin
    LATBbits.LATB15=1;

    // get ADC value
    ADCValue = (ADC1BUF0"2); // yes then get ADC value

    // Relaunch ADC sampling
    AD1CON1bits.SAMP = 1; // start sampling

    // Update DDS value
    phase+ = phaseincrement;                  // Increment the
    phase register
    lookupindex = (unsigned char)(phase"8); // get the top 8
      bits
    DDSValue = SINE_TABLE[lookupindex];     // and convert it
      to a sine

    // Do the numerical mixing
    MixedValue = (signed char)((((signed short)DDSValue-
      128)*((signed short)ADCValue-128))"8);

    // Integrate, first step
    integrate1+ = MixedValue;
```

```
// Integrate, second step
integrate2+ = `integrate1;

// Time to decimate ?
if(CombCpt++> = COMBRATIO-1)
{
    CombCpt = 0;

    // The following code is executed only at 1/16th the rate
    // it could be executed by a lower priority task if
       needed
    comb1 = integrate2-lastint2;
    comb2 = comb1-lastcomb1;
    lastint2 = integrate2;
    lastcomb1 = comb1;
}

FilteredValue = (signed char)(comb2"8);

// Put it to output port, depending on selected output
    switch(PORTA&0x0F)
{
    case 0 :
        LATB = (LATB&0xFF00)|ADCValue;
        break;
    case 1 :
        LATB = (LATB&0xFF00)|DDSValue;
        break;
    case 2 :
        LATB = (LATB&0xFF00)|(MixedValue+128);
        break;
    case 3 :
        LATB = (LATB&0xFF00)|((FilteredValue"2)+128);
        break;
    default :
        LATB = (LATB&0xFF00);
}

// Signal end of interrupt processing on B15
LATBbits.LATB15=0;
}
```

```
// Main program

int main(void)
{
    // Port B is output, initialized to 0
    TRISB = 0x00;
    LATB = 0;

    // initialize DDS
        Phase = 0;
        Phaseincrement = 18481; // Fout = 25KHz x
        phaseincrement/65536 = 7.050KHz

    // Configure ADC
    AD1PCFG = 0xEFFF; // all PORTB = Digital; RB12 = analog
    AD1CON1 = 0x00E0; // SSRC<3:0> = 111 implies internal
    // counter ends sampling and starts converting.
    AD1CHS = 0x000C; // Connect AN12 as CH0 input.
    // in this example AN12 is the input
    AD1CSSL = 0;
    AD1CON3 = 0x1F02; // Sample time = 31Tad,
    // Tad = 2 Tcy
    AD1CON2 = 0;
    AD1CON1bits.ADON = 1; // turn ADC ON
    AD1CON1bits.SAMP = 1; // start sampling

    // Configure PORTD to output, 0
    PORTB = 0;
        TRISB = 0;

    // Configure timer2 for periodic interrupt
        ConfigIntTimer2(T2_INT_ON|T2_INT_PRIOR_1);/*Enable
          Interrupt*/
        OpenTimer2(T2_ON,0x9C40/250); //Timer is configured for 40 usec
    // Main loop
    while(1);
}
```

I have not actually built this project, but thanks to Labcenter's VSM mixed-signal simulator I was able to test it on my PC. Have a look at Figure 8.9. The beauty of a tool like VSM is that it allows you to simulate the analog portion of the design like any

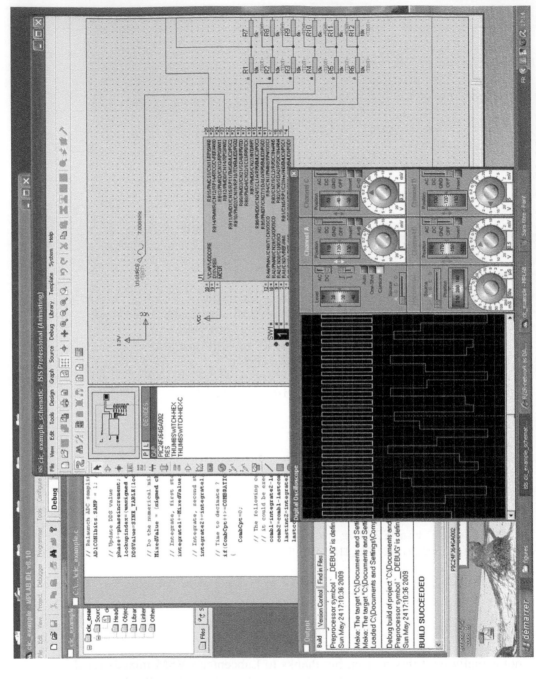

Figure 8.9: Labcenter's VSM allows both the hardware and the embedded firmware to be simulated. You can see MPLAB IDE on the left and VSM on the right, with the virtual scope in action.

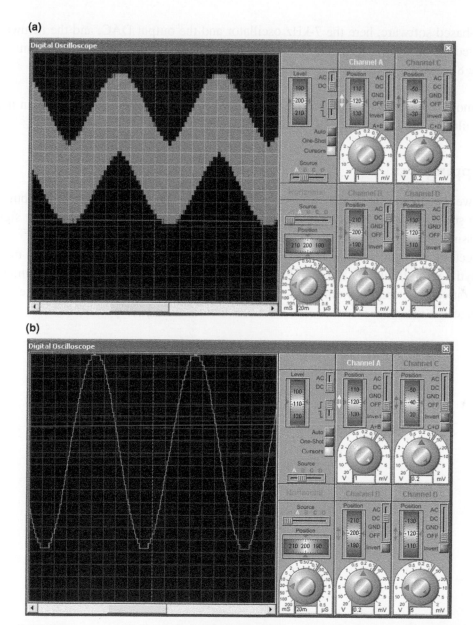

Figure 8.10: The output of the digital mixer (a) shows, as expected, a 50-Hz signal added with a high-frequency noise, which is nothing else than the 14-kHz image frequency. When the CIC filter (b) is put in action, the virtual scope shows a very clean 50-Hz sine.

Spice-based software, here the 7-kHz oscillator and the output DAC, and the firmware executed inside the PIC. The result is very close to expectations that are shown in Figure 8.10 on page 135.

I am more than confident that this design would have worked with no issues even if I had built it using the old method, namely a soldering iron.

Wrapping Up

Here we are. In summary, CIC filters are actually simple moving average filters but with a smart implementation. These filters will allow you to break down difficult signal processing problems and drastically reduce the requirements on the computing architecture, especially when combined with software-based DDS and frequency mixers. So anytime you think that the useful information is far smaller than the actual sampling rate, you can be sure that multirate techniques can help. And now these techniques are no longer on the darker side, so you have one more tool in your pocket!

Part 4
Oscillators

CHAPTER 9

Let's Be Crystal Clear

I still consider myself reasonably young, but I remember when my dad bought one of the first inexpensive electronic watches. The display was made of 7-segment red LEDs, so I had to press a button to read it. Although the batteries drained after a couple of weeks, it was still a great innovation. As you know, such a watch is based on a crystal.

Crystals remain important. Many of the electronics that you use daily, from MP3 players to wireless devices, probably wouldn't be possible without the time stability of crystals. Although crystals are very common, they are often misunderstood. For example, do you know why there are usually small capacitors around a crystal? Do you understand the difference between a "series" and a "parallel" crystal? In this chapter, I will try to answer these questions and a couple of others. Follow me into the magical world of crystal oscillators.

Piezoelectricity

Some crystals, quartz in particular, are piezoelectric devices. Piezoelectricity is on the border between electricity and mechanics: If you press a piezoelectric device, a voltage will appear between its sides. Conversely, if you apply a voltage between two of its opposed faces, its physical size will change. Piezoelectricity was first demonstrated and explained in Paris in 1880 by Pierre and Jacques Curie. (The former became the husband of Marie Curie, and they later discovered small things such as radioactivity,

Note: This chapter is an improved reprint of the article "The darker side: Let's be crystal clear," *Circuit Cellar*, no. 215, June 2008.

DOI: 10.1016/C2009-0-20196-6

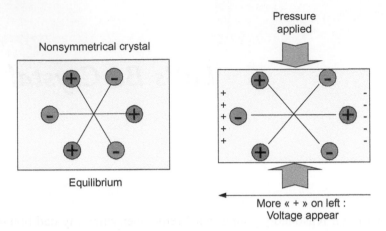

Figure 9.1: Piezoelectricity appears when a crystal is not fully symmetrical. Look at the simplified drawing on the left. Without any pressure, these atoms are centered and the overall electrical field is null. If you press vertically on the crystal, you will slightly change its shape (right), the electrical charges will no longer be symmetrical, the left side will be a little more positive than the right, and a voltage will appear. Conversely, if you apply a voltage, the crystal will distort itself.

but that's another story.) The underlying physical principle is quite simple (see Figure 9.1).

As usual, the first real-life application of a crystal was for the military. Paul Langevin used it as a transducer to create ultrasound for the first sonar system during World War I. According to the *Encyclopedia Universalis*, the first quartz-based oscillator was built in 1918 by Walter G. Cady, and that was the first of many. During World War II, the United States built more than 50 million quartz-based oscillators for its transmitters!

The piezoelectric phenomenon is at microscopic scales. Unfortunately, you will never see a crystal size change when it is electrically excited. A few volts on a quartz crystal correspond to an electricity charge of a couple of picocoulombs, which translates into forces in the Newton range (100 g), which translates into displacements of a few micrometers only. Quartz manufacturers know how to cut a crystal to optimize its performance for a given application. The cuts are normalized in regard to the crystal structure (named X-cut, Y-cut, CT-cut, AT-cut, etc.). But you usually won't need to go into this level of detail because you are on the electrical side.

The Electrical Model

Quartz shrinks or inflates depending on the voltage. So if you apply an AC voltage with a specific frequency to the two terminals of a crystal, you will enable it to reach its mechanical resonance (either longitudinal vibration resonance or a more complex shearing resonance depending on the crystal cut). At that precise frequency, the current flowing through the crystal will be in phase with the excitation signal and the crystal will electrically behave as a small resistor—let's note it $R1$. If you move the frequency around the resonance frequency, the crystal will behave as a series $R1/L1/C1$ network, in parallel with a loading capacitor $C2$. $C1$ is called the motional capacitance, and it models the behavior of the quartz at resonance with $L1$. $C2$ is nothing more than the parasitic capacitance between the two terminals of the quartz (see Figure 9.2).

The values of the equivalent components make quartz a very uncommon *RLC* network. Most megahertz-range crystals have an inductance (L), not in nanohenries but from 0.1 to 1 H. The series resistance (R) is hundreds of ohms, and the motional capacitance (C) is in femtofarads (yes, 0.001 pF), providing a quality factor of more than 100,000. Because the quality factor of a network is equal to the bandwidth divided by the central frequency, this helps to build precise and stable oscillators. It also explains why you can't replace quartz with equivalent discrete components. Just try to buy a 1-H inductor with a parasitic capacitance below 1 picofarad. Finally, the parasitic capacitor $C2$ is a more usual value, a couple of picofarads.

For frequencies far from the resonant frequency, the crystal is equivalent to the capacitor $C2$, because the impedance of $C1/L1/R1$ is high. However, when you start low and increase the frequency, you reach the resonant frequency of $L1$ and $C1$ (noted F_{SERIES}) at some point. At this frequency, the crystal will behave like a simple low-value resistor $R1$: This is its minimum impedance. If you continue to increase the frequency, the crystal will behave like a larger inductor, up to a virtual open circuit. It will finally reach the behavior of the capacitor $C2$ (see Figure 9.2).

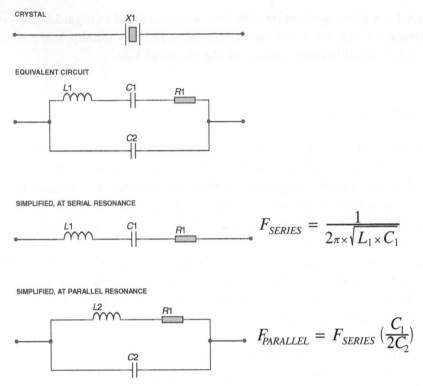

Figure 9.2: The equivalent circuit of a crystal is a series *L1/C1/R1* resonant circuit in parallel with a parasitic capacitance *C2*. Such a network has minimum impedance when the *L1/C1* network is at resonance (series resonance), and at maximum impedance when *L1/C1* behaves as a large inductance *L2*, which oscillates with the parasitic capacitance *C2* (parallel resonance).

It's easier to see it rather than describe it, so I hooked my Hewlett-Packard 3585 network analyzer on a standard 8-MHz crystal. The plot is shown in Figure 9.3. You can do the same with a good, stable sweeper generator and an oscilloscope, but an actual network analyzer makes life easier.

As you can see, my test crystal shows a clear series resonance (minimal impedance) at 8.000250 MHz. It then has a maximum impedance at a frequency around 12 kHz

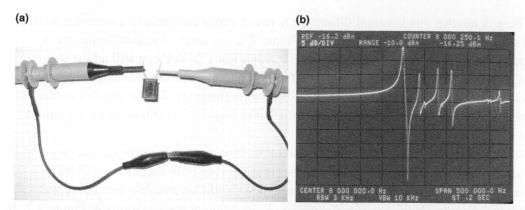

Figure 9.3: This is the transmission response of an 8-MHz crystal oscillator (a) as grabbed on a Hewlett-Packard 3585 spectrum/network analyzer (b). The minimum impedance corresponds to the maximum transmission, here at 8,000,250 Hz. This point is followed by a minimal transmission (maximal impedance) that is a couple of kilohertz higher, and then several spurious resonances are visible.

higher. At some tens of kilohertz higher, the crystal shows other resonances but with lower amplitudes. These resonances are parasitic resonances generated by other vibration modes of the crystal often linked to mechanical discontinuities. They usually can be disregarded because your oscillator will be designed to lock on the main resonances.

Parallel or Series?

This simple experiment highlights an extremely important concept: *There are two closely spaced usable resonance frequencies for a given crystal.* The first one, the series resonant frequency, F_{SERIES}, is the intrinsic resonance frequency of the crystal $L1/C1$ and corresponds to minimal impedance. The second resonant frequency, $F_{PARALLEL}$, is a little higher in frequency and corresponds to maximum impedance. This is the resonance of the $R1/L1/C1$ network, which behaves here as a high-value inductor, close to an open circuit, with the parasitic capacitor $C2$ (see Figure 9.2 again). There isn't a "series cut" or "parallel cut" crystal; all crystals are the same. However, a crystal built to be used at its parallel resonant frequency is adjusted a little below to compensate for the small difference between F_{SERIES} and $F_{PARALLEL}$.

There is another fundamental difference between series and parallel resonance modes. As shown in the formulae provided in Figure 9.2, the series resonant frequency is independent from the value of the parasitic capacitor $C2$. You can still adjust the frequency of a series-made crystal oscillator by adding series inductors or capacitors, which will slightly change $L1$ or $C1$, respectively. However, for the parallel mode, the value of the parasitic capacitor $C2$ is influencing $F_{PARALLEL}$, as shown in the equations.

Moreover, because this capacitor is between the two terminals of the quartz, any external loading capacitor is also influencing it because both capacitors will be in parallel. Thus, the parallel oscillation frequency should be unstable depending on external circuits. However, there is a trick. Look again at the equations in Figure 9.2. If $C2$ becomes higher and higher, then the $F_{PARALLEL}$ frequency becomes closer and closer to F_{SERIES}. It is easy to show this with the network analyzer. Refer to Figure 9.4.

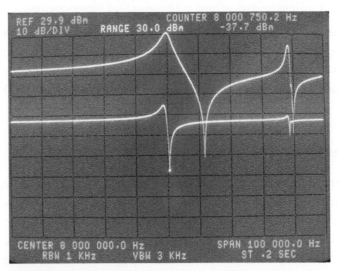

Figure 9.4: The addition of a 20-pF capacitor in parallel to the crystal (*bottom curve*) doesn't change the frequency of the series resonance compared with the quartz alone (*top curve*). However, the parallel resonance is now far closer to the series resonance—that is, 500 Hz away compared with 12 kHz. This is clear with the formula provided in Figure 9.2. If C2 gets higher, both frequencies get closer.

The addition of a 20-pF capacitor in parallel with the crystal doesn't change the series resonant frequency. However, the parallel resonance is reduced from 12 kHz to 500 Hz away from F_{SERIES}, making it more resistant to external variations. *That is why a crystal in parallel mode is always used with an external parallel loading capacitor.* And that's why a crystal manufacturer always indicates the value of the external capacitor if the given crystal is designed to be used in parallel mode in its specifications. If you load it with a capacitor of a different value, your oscillator simply won't oscillate exactly at the value stamped on the crystal body.

By the way, the experimental setup used for Figure 9.4 is also a good way to calculate the $C1$ motional capacitor of a crystal: Measure the frequency variation of the parallel resonance with a given external capacitor and deduce $C1$ from the equation in Figure 9.2. This is called the pullability method.

HCMOS Pierce Oscillator

Now let's examine a classic quartz-based oscillator. Nearly all CMOS-based oscillators use parallel resonance. Why? Because a series mode means very low impedances and CMOS oscillators are happier with large impedances. Moreover, series resonance is more sensitive to temperature and component variations. I built an oscillator on a prototyping board around an 8-MHz crystal and a pair of 74HC00 gates (see Figure 9.5).

Let's go through each component around the crystal. First, the ubiquitous $C1$ and $C2$. The two capacitors form the external loading capacitor I just described. Rather than use a single capacitor across the terminals of the crystal, it is more stable to use two capacitors each between a terminal and the ground. The two capacitors act as a single capacitor because they are connected in series, at least from the point of view of the crystal, and stabilize the oscillation more efficiently than a single floating capacitor. Their value must be twice that of the loading capacitor that is specified for the crystal you are using, usually around 20 pF each, because the overall capacitance of two C capacitors in series is $C/2$. The value of the capacitors must also be trimmed to account for all of the parasitic capacitors because you need to get exactly 8 MHz.

Because you are in parallel resonance mode, the crystal will show maximal impedance and a phase shift around 90°. The HCMOS inverter provides a phase shift around 180°,

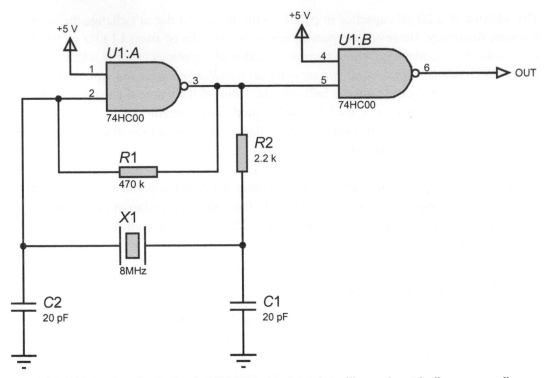

Figure 9.5: Here is the basic HCMOS gate-based oscillator that I built on a small protoboard for the experiments provided in this chapter. A parallel mode crystal oscillator can't be simpler.

plus some propagation delay through the HCMOS gate. An oscillating loop must provide a 360° phase shift through the loop, so a little more phase shift should be added. The resistor *R2* helps here. It adapts the input impedance of the crystal to a higher value and then shifts its operating point to a frequency where its phase shift is 180°. *R2* also helps reduce the risk of spurious oscillations. As a starting point, the value of the resistor should be close to the impedance of the loading capacitor at the working frequency $(1/(2\pi \times F \times C)$, giving around 2 kohms. The resistor also avoids loading the output of the logic gate with a strange load as a crystal directly connected would.

Finally, the resistor *R1* is just helping to keep the NAND gate in its linear region: The logic gate is used as an inverting analog amplifier and not as a pure logic gate. The

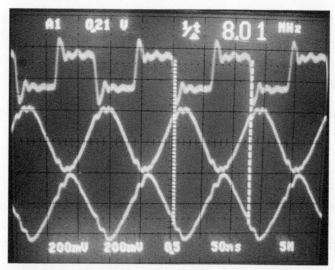

Figure 9.6: This screenshot is a bit blurry, but I think you'll still find it useful. The top curve is the output of the oscillator (8 MHz as planned). The two bottom curves show signals measured on both ends of the crystal. They are clearly 180° out of phase or so, as expected.

value of $R1$ is not critical, but it should be high enough to reduce power requirements (because the input impedance of a CMOS gate is very high), usually between 100 kohms and a couple of megaohms.

Figure 9.6 shows the waveforms measured on the output of this oscillator as well as on the output of the two terminals of the crystal. As expected, there is nearly a 180° phase shift between the two terminals of the crystal. The output frequency was measured at 8.0021348 MHz. Why not exactly 8 MHz? I have wired this oscillator on a protoboard, so long wires are connecting the crystal and the logic gates, adding parasitic capacitors and inductors everywhere.

Starting from this basic schematic, what happens if you change a component's value? As expected, increasing $C1$ or $C2$ decreases the oscillating frequency. For example, adding 10 pF reduces the frequency to 8.0016469 MHz. The resistor $R2$ also has an effect on the frequency. Increasing it to 10 kohms reduces the oscillating frequency to 8.0017738 MHz. However, this change degrades the stability of the output. The standard deviation of the output frequency increases from 0.249 to 1.678 Hz, but this is more difficult to see without an exotic measurement tool. I will cover this later.

Oscillator Start-Up

It is also interesting to check the behavior of the oscillator when its power supply is switched on. It is easy to do. Simply hook an oscilloscope on its output, trigger it on the raising front of the power supply, and, voila, you have the screenshot shown in Figure 9.7.

The start-up behavior is a little chaotic, but after around 12 μs the output signal amplitude is stabilized. So you will conclude that the start-up time of the oscillator is 12 μs. You are wrong, but it is not easy to see why with an oscilloscope. I am a lucky

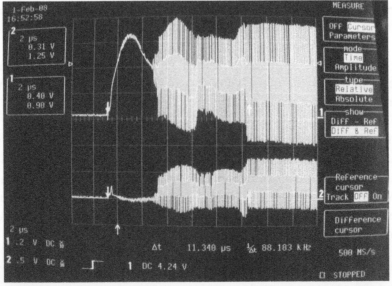

Figure 9.7: If you capture the start-up behavior of my test crystal oscillator with an oscilloscope, you may conclude that it is stabilized after about 12 μs. This is true regarding amplitude but not frequency. The horizontal scale is 2 μs/division, the *top* curve is the quartz output, and the *bottom* is the logic gate output.

engineer. My consulting company has some pretty cool test equipment. In particular, we have a Hewlett-Packard 5372A frequency and time interval analyzer (see Figure 9.8a). The unusual box is a powerful universal counter. It can obtain frequency or time measurements quickly, up to millions of measurements per second (and display either their variation over time or their statistics). Let's connect it to the oscillator output and trigger it when power is switched on.

The result is the plot in Figure 9.8b, which shows the output frequency of the oscillator on the vertical axis and the time on the horizontal axis. At the right of the graph, around

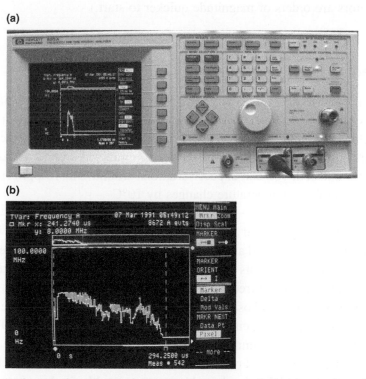

(a)

(b)

Figure 9.8: (a) The Hewlett-Packard 5372A frequency and time interval analyzer enables you to capture the frequency behavior of the crystal at start-up. (b) Here the frequency first jumps to 40 MHz or so and goes down and locks to 8 MHz after only 241 μs (marker position). Time is on the horizontal axis, and frequency is on the vertical axis from 0 to 100 MHz. A nice instrument, isn't it?

240 µs after start-up, the oscillator is well settled at 8.000 MHz, but during these 240 µs the oscillator output goes anywhere from 0 to 40 MHz! This is the intrinsic stabilization time of the oscillator.

The conclusion is that you shouldn't use the output of a crystal oscillator during a couple hundred microseconds. And this is why microcontrollers usually include oscillator start-up timers, just to be sure not to over-clock the silicon during this stabilization time. The oscillator design can be optimized with a good selection of external components, but this mandatory delay will never be a couple of microseconds. Be prepared to wait, or use something other than a crystal. (*RC* oscillators or even ceramic resonators are orders of magnitude quicker to start.)

Is It Stable?

Now the crystal oscillator is started and stays on 8.000 MHz. You know that this is a little false, don't you? First, its actual frequency will drift with the ambient temperature. The shape of the temperature-to-frequency curve is not the same for different crystal cuts, but manufacturers ensure that the variation is minimized around the ambient temperature. In a nutshell, you can expect variations around 1 to 10 ppm (parts per million) when the ambient temperature changes by 10°C.

There are two solutions if you need a far more stable oscillator. The first one is to add a circuit that will measure the actual ambient temperature, and slightly tune the quartz oscillator with external components such as varicap diodes, based on some knowledge of the theoretical temperature-to-frequency error. Such oscillators, which are called temperature-compensated crystal oscillators (TCXOs), will provide stabilities of 0.1 to 0.5 ppm per 10°C. The correcting circuitry can be either analog or digital. (DTCXO is a good idea for a future project around your preferred microcontroller.) By the way, you can find inexpensive TCXO-integrated products if you need a low-frequency oscillator such as a standard 32-kHz-based RTC. Look at chips such as the Maxim Integrated Products DS32B35, which provides a 2-ppm stability from 0 to 40°C for a couple of bucks.

If you need an even greater stability over temperature, you must pay more for oven-controlled crystal oscillators. The idea is simple. Heat the crystal to a fixed temperature

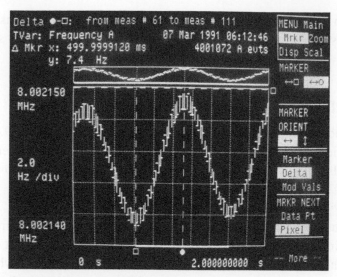

Figure 9.9: This is what happens if a 50-mV/1-Hz "noise" is added to a 5-V power supply: The frequency of the crystal oscillator is gently following the same shape as the power supply voltage. Don't be confused by the graph. This is not a time domain oscillograph but the output frequency of the oscillator over time. Each vertical division is a 2-Hz variation.

so it will no longer drift if the ambient temperature changes. That is easy to say, but it is more difficult to do, especially for power- or cost-restricted applications.

But temperature is not the only factor that can impact your frequency. Do you want another cool experiment? I've added a little noise on a clean 5-V powering logic gates–based 8-MHz oscillator. That is not a lot of noise, just a 50 mVpp, 1-Hz sine added to the 5 V, so a power supply moving slowly from 4.95 to 5.05 V. Is there a visible impact on the output frequency? Time to switch on my 5372A analyzer again. It generates what you see in Figure 9.9.

The output frequency closely follows the power voltage sine curve, with a ±3-Hz variation! Not a large error, but it can be a problem for some applications. Temperature and supply voltage are usually the first contributors to oscillator problems, but they are not alone. Remember that quartz is a piezoelectric device. If you stress it mechanically (e.g., with vibrations), there can be a visible change in the output frequency. On some

RF designs, even sound waves can generate visible spurious modulations of the output (the so-called nasty microphonic effect), so high-g designs usually have some concerns with crystals.

Let's Overtone

Finally, the impedance of a crystal over frequency, as shown in Figure 9.4, gives a beautiful resonance. But what happens at higher frequencies? The answer is in Figure 9.10: A crystal also has overtone resonance frequencies, usually close to three, five, or even seven times its basic frequencies.

It is possible to use a crystal in such an overtone mode to build higher-frequency oscillators. It is important to understand that the overtone modes are not electrical harmonics of either F_{SERIES} or $F_{PARALLEL}$; they are mechanical overtones where the crystal is oscillating in more complex mechanical ways. The consequence is that the third overtone frequency is not exactly three times higher than its base frequency, even if it is

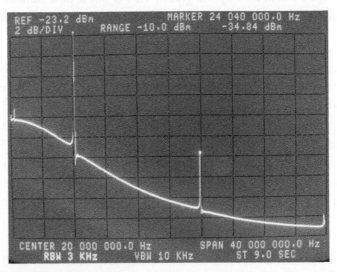

Figure 9.10: A crystal can also be used in overtone resonance modes. Using the same test setup as the one used for Figure 9.3, this is what happens when the frequency is significantly higher than the crystal's fundamental frequency. A third-order resonance is clearly visible around 24 MHz, three times 8 MHz.

close. That's also why you can buy crystals specifically built to operate in overtone. First, their construction is optimized to enhance the overtone resonances, but they are also tuned to provide the exact desired frequency in the overtone mode.

Building an overtone oscillator is theoretically simple, but it is usually a little tricky on the bench. Let's look at the theory first. Just add a bandpass filter circuit in series with the crystal that will damp the fundamental frequency. Refer to Figure 9.11, where my dear logic-gated oscillator is modified to oscillate on the third overtone (more or less $3 \times 8 = 24$ MHz).

The $C1/L1$ network acts as a resonant bandpass filter with a center frequency of 25 MHz (i.e., $1/(2\pi \times \sqrt{(L1 \times C1)})$). The actual measured frequency is 23.95966 MHz,

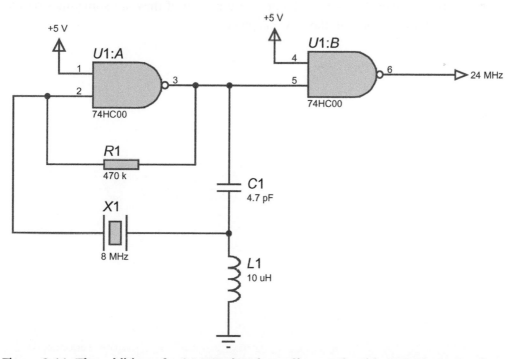

Figure 9.11: The addition of a 24-MHz bandpass filter made with *C1* and *L1* transforms the 8-MHz oscillator shown in Figure 9.5 into a 24-MHz overtone oscillator.

which is, as expected, close but not equal to three times the fundamental frequency. What makes an overtone oscillator tricky is the parasitic components, which are increasingly difficult to manage. Here I had to remove the 20-pF capacitors to help the oscillator oscillate where I wanted, as the parasitic components started to be significant at such a high frequency. Anyway, with a couple of experiments you should get it right.

Wrapping Up

I hope that your ideas about crystals, parallel versus series modes, overtones, and so on, are now clearer. You should understand why you shouldn't forget to put two capacitors around your crystal, as shown in Figure 9.4; why you should minimize wirings; and why you should have a clean power supply. Don't hesitate to read through the documents and articles listed in the References. They will provide you with additional details about these topics. Crystal oscillators are fun, even if they are sometimes on the darker side. We will use them in the next chapters!

CHAPTER 10

Are You Locked?
A PLL Primer

In this chapter, I will dig into one of the most useful electronics building blocks, the phase-locked loop, nicknamed PLL. I don't know why, but a significant number of electrical engineers are frightened when someone pronounces this acronym. This may be because PLLs are sometimes presented in textbooks through their mathematical properties rather than through their silicon side. That is a shame because PLLs are actually quite simple and they are definitely useful! Let's try to prove it.

A PLL enables you to generate an output frequency based on a reference input clock. The output frequency can be either higher or lower than the input, but the fundamental point is that the PLLs make sure that they stay locked to each other. That means that the precision of the output and the input will be the same. If you have a high-stability, 10-MHz, 2-ppm (part per million) clock, you will be able to generate virtually any frequency with 2-ppm precision. Moreover, PLLs enable you to "clean" a noisy reference clock, filtering any short-term noise or jitter while keeping the long-term stability of the reference, thanks to the locked loop (see Figure 10.1).

With these properties, it is not surprising to find PLLs in virtually all RF devices. Transmitters need PLLs to generate and modulate the carrier frequency, and receivers need them to generate the local oscillators and to recover the bit rate from a noisy signal. PLLs are also present in a large number of digital systems. You probably meet a PLL every time you open a device that needs more than one clock source, in variable clocks, or in devices that need to work with noisy digital signals. Think about it: That

Note: This chapter is an improved reprint of the article "The darker side: Are you locked? A PLL primer," *Circuit Cellar*, no. 209, December 2007.

© Elsevier Inc.
DOI: 10.1016/C2009-0-20196-6

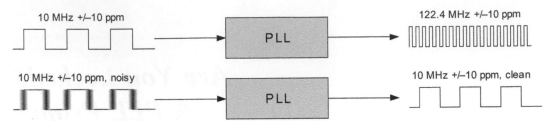

Figure 10.1: The first classic application of a PLL is to generate a precise frequency (here 122.4 MHz) based on another precise reference frequency (10 MHz). Another important use of PLLs is to clean a noisy reference frequency, removing any short-term phase variations while keeping the same long-term precision.

means everywhere. Some examples include Ethernet interfaces, multirate MP3 players, DVD readers, pixel clock recovery for TV or cellular receivers, and, of course, your PC. Without PLLs, multi-gigahertz processors would not exist.

VCO Basics

One of the main components of a PLL is a voltage-controlled oscillator (VCO). You will usually not have to build your VCOs yourself, but understanding how things work doesn't hurt, so let's spend some time on this interesting subject. A VCO is a special kind of oscillator. Building an oscillator is not a difficult task; a badly designed amplifier will naturally oscillate, especially if it is not the designer's intention. Building a good oscillator is, of course, a little more complex, but any amplifying circuit—meaning one with a gain that is greater than 1—will oscillate when its output is fed back on its input, intentionally or not. However, there is an additional condition. There must be a given frequency where the amplifier phase shift is 0. At that frequency, the output signal will be in phase with the input signal. The frequency will be the oscillation frequency.

Figure 10.2 shows a basic common emitter amplifier stage preceded by a filtering stage (as well as its simulated gain and phase response). I did this simulation with the Proteus CAD tool suite from Labcenter Electronics (UK), which provides a user-friendly front end for the well-known Spice simulation engine.

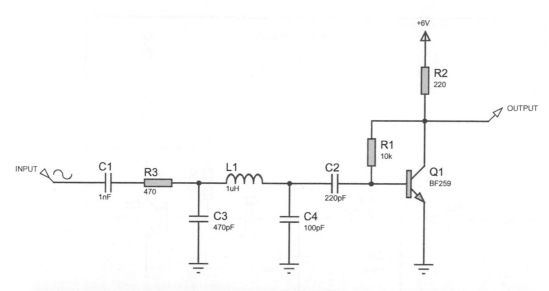

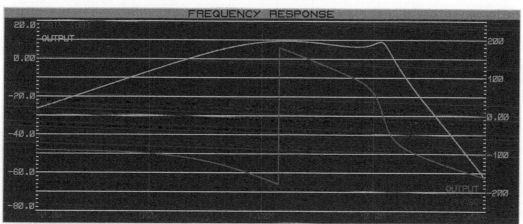

Figure 10.2: This simple transistor-based amplifier has a positive gain from 250 kHz to 20 MHz (*top*). The phase response (*bottom*) indicates that its phase response crosses the 0 line at a frequency of 13 MHz.

The simulation shows that the gain of the amplifier is greater than 1 (above 0 dB) from roughly 250 kHz to 20 MHz. Its phase response is 0 only at a given frequency, 13 MHz. What happens if you connect its output to its input? You bet it will oscillate at around 13 MHz (see Figure 10.3).

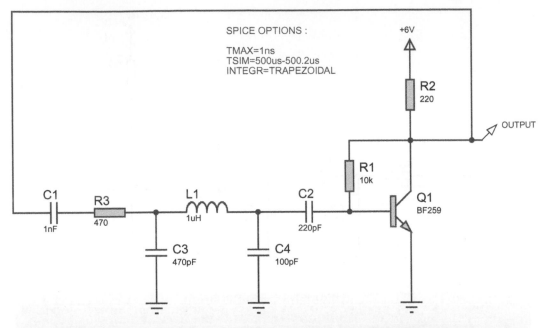

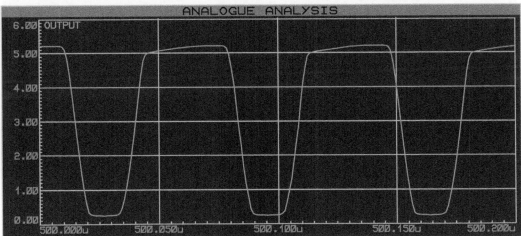

Figure 10.3: This simulation shows you that if you take the schematic in Figure 10.2 and connect its output to its input, you have an oscillator. The oscillation frequency is 14.6 MHz, close to the frequency where we got a null phase response in the open-loop configuration.

The simulated oscillation frequency is 14.6 MHz. This is due to a change in the amplifier response because we are now loading its output and capacitively loading its input differently from the open-loop case. A word of caution here: Simulating an oscillator is not always easy with Spice (or whatever tool you are using) because an oscillator needs some noise to start. This is easy in real life but more difficult in the memory of a computer. The usual tricks are to force a short time step in the Spice parameters (1 ns) and declare that one of the capacitors must be precharged at the start of the simulation. You must also visualize the simulation result some time after the origin in order to let the oscillator stabilize (from 500 to 500.2 µs in the examples). Labcenter Electronics's support was a great help in pinpointing the tricks.

Converting an oscillator into a VCO is not difficult. Because the oscillation frequency is dependent on the frequency-response curve of the underlying amplifier, you have to make the response voltage dependent. This is usually done with a varicap diode because a diode presents a variable capacitance depending on the inverse DC voltage applied to its pins. Figure 10.4 shows the phase response of a slightly modified version of our amplifier design, with a varicap diode polarized by a DC voltage from 0 to 25 V. This voltage is generated thanks to the R4/R5 divider and applied through the L2 inductance, which behaves as an open circuit at high frequencies. The 0 phase response is moved from 11.4 to 12.3 MHz when the voltage is modified. If we loop back the circuit, we have a VCO.

The schematics in Figure 10.4 are simplified and designed to illustrate the concept, but typical VCOs built around Pierce or Colpitts oscillator topologies are not very different. You can find good prebuilt VCOs from suppliers such as Mini-Circuits, Synergy Microwave, Sirenza Microdevices, and Z-Communications. However, knowing how a VCO is built is always useful. The output frequency of a VCO is driven by a DC voltage, so you can easily deduce that any noise on the DC input voltage or any thermal noise in the VCO components themselves will produce a "frequency noise" on the output signal. This means that the signal's frequency will not stay exactly the same from cycle to cycle. This kind of noise is usually called phase noise, and it is specified in an exotic unit, the dBc/Hz (see Figure 10.5).

Phase noise can also be understood as jitter in the time domain, even if the phase-noise–to–jitter relationship is not straightforward mathematically speaking. If you are interested in this mathematical relationship, you can find good information in Brad

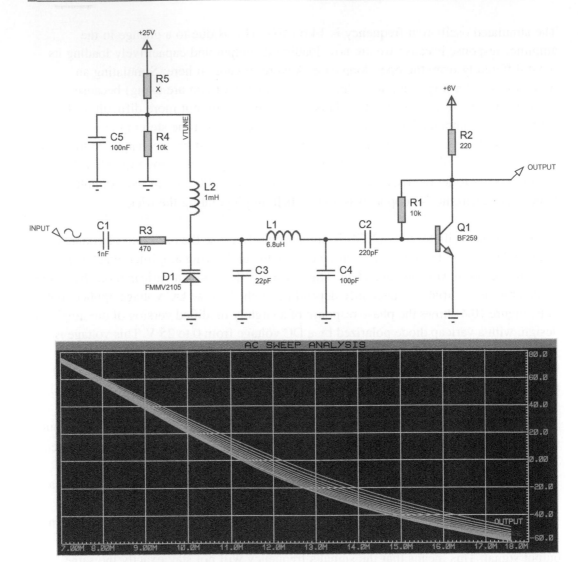

Figure 10.4: To transform an oscillator into a voltage-controlled oscillator, you just need to add a varicap diode because its capacitance will change depending on the applied DC voltage, which is simulated here by a variation of the R5 resistor value. The design will give a VCO a 1-MHz tuning range at around 13 MHz.

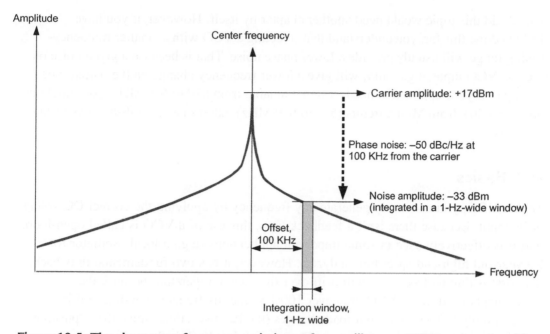

Figure 10.5: The short-term frequency variations of an oscillator or VCO are measured in dBc/Hz. This strange unit is just a convenient and normalized way to measure how quickly the noise power drops when you examine frequencies farther from the carrier. On this spectrum analyzer plot, the power of the signal is 17 dBm or 50 mW (i.e., 1 mW × 70/10) because a "dBm" is a decibel relative to 1 mW by definition. Due to noise, the power measured 100 kHz away through a hypothetical 1-Hz-wide bandpass filter is not null but −37 dBm. This is 54 dB (i.e., 17 dB − 37 dB) lower than the carrier. Because we are comparing the carrier power, the unit is "dBc," meaning "dB relative to the carrier," and "dBc/Hz" as we normalize to a 1-Hz-wide noise window. If we measure with a 10-Hz-wide filter, the noise power will simply be 10 times higher.

Brannon's application notes on the Analog Devices Web site (see References). Phase noise is often the main difficulty a VCO user must fight against. For example, if you design a transmitter, a high level of noise will give you a noisy one. And if you design a receiver, you will end up with a poorly selective device. Phase noise can be a great concern even in digital applications. For example, if the VCO is used as the clock source for a high-speed ADC, its phase noise (i.e., its jitter) must be strictly controlled in order to achieve a good signal-to-noise ratio.

I'm afraid this topic would need another chapter by itself. However, if you have followed me this far, you understand that selecting a VCO with a smaller frequency-tuning range will usually provide a lower phase noise. That is because a given noise on the DC VCO input, e.g., 1 mV, will give a lower frequency change on the output. For example, if your application needs a signal tunable from 610 to 620 MHz, you should use an ROS-630+ from Mini-Circuits (595 to 630 MHz) rather than a JTOS-850VW (400 to 850 MHz).

PLL Basics

With a VCO, you can easily generate any frequency by applying the correct DC voltage at its input. Because there isn't a feedback loop, this use of a VCO is called open-loop, and it is effectively used in some important applications (e.g., a local oscillator for a 20-year-old high-end spectrum analyzer). However, it has two fundamental drawbacks. The first is that its frequency will drift over time and temperature because the components used in the VCO are never 100% stable. Its frequency will also drift if the power supply voltage or output load changes (the last two effects are called "pushing" and "pulling" in VCO specifications). The second drawback is that any voltage noise on the DC control input will generate frequency domain noise, which means bad phase noise characteristics.

Using a PLL is an efficient way to wrap an imperfect VCO in a circuit in order to lock its output frequency on a more stable reference frequency, as well as to reduce its phase noise. PLLs date back to the 1930s, when the first homodyne RF receivers were designed. Homodyne receivers are zero intermediate frequency receivers, mixing the antenna signal with an oscillator tuned exactly to the transmitter carrier frequency. Because the local oscillators were unstable, a way to constantly retune them was needed. British engineers built an automatic correction loop based on a measurement of the actual receiver output. The design, based on an idea published in 1932 by the French scientist H. de Bellescise, was probably the first PLL. The overall architecture of a PLL can be seen in Figure 10.6.

The key idea is simply to compare a reference frequency (F_{IN} divided by a constant R if needed) and the actual output frequency of a VCO (F_{OUT} divided by a constant N if needed), using a block called a phase frequency detector (PFD). The error signal is then

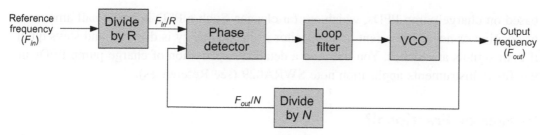

Figure 10.6: The architecture of a PLL is built around a phase detector that drives a VCO through a filtering stage. The output of the VCO is then looped back to the phase detector and compared to a reference frequency. Dividers enable you to change the frequency ratio between the reference and the output. So-called fractional-N PLLs have the same architecture but allow non-integer dividers.

filtered and used to drive the VCO frequency control input. How does it work? The loop is stabilized when both phase detector inputs have equal frequencies (and are in phase), meaning that F_{IN}/R is equal to F_{OUT}/N. Simply multiplying both sides of the equation by N gives

$$F_{OUT} = \frac{N \times F_{IN}}{R}$$

If the VCO drifts and its frequency increases, the phase detector will see the change and decrease its output voltage, which will then reduce the VCO frequency until it is back to being equal to $N \times F_{IN}/R$. The same will happen if the drift is in the other direction. The frequency output of the VCO will stay locked on the value $N \times F_{IN}/R$ thanks to the loop. That's why we call it a PLL. The frequency at which the phase detector is working (F_{IN}/R is equal to F_{OUT}/N) is called the PFD frequency and is a fundamental choice in the design of a PLL.

I already covered the architecture of a VCO, and I'm sure you know that a digital frequency divider is easily built with a binary counter. The other building block of a PLL is the PFD. The first PLL chips used an exclusive OR gate as the detector. If both inputs are frequency and phase locked, the gate provides a stable rectangular signal, which is transformed into an average DC value with a good low-pass filter, which can then drive the VCO. The only issue is that, because the filter is not perfect, the PFD frequency "leaks" to the VCO input, which gives a noisy output. Modern PLL chips are based on charge pump PFDs, which are far cleaner. Such a PFD injects small amounts of positive or negative current in a capacitor if the frequency is drifting, and stays quiet if both signals are locked. You will find a detailed explanation of charge pump PFDs in the Texas Instruments application note SWRA029 (see References).

Integer or Fractional?

The PLL architecture in Figure 10.6 is called an integer-N PLL. It is a simple and elegant architecture, and you can use it to generate any frequency that can be expressed as $N \times F_{IN}/R$. Suppose that you have a 10-MHz reference frequency and you want to be able to tune the PLL output to around 500 MHz with a resolution of 1 kHz. The 1 kHz is called the channel spacing. Then the solution will be to set the R divider to 10,000 (i.e., 10 MHz/1 kHz). F_{OUT} is $N \times F_{IN}/R$, which translates to $N \times 1$ kHz. So the PLL will generate a 500-MHz signal if you set $N = 500,000$, or a 500-MHz + 1-kHz signal with $N = 500,001$, and so on. *With an integer-N PLL, the rule is that the PFD frequency must be equal to the desired channel spacing.* The required channel is then selected with the N divider.

Integer-N PLL chips are used in many products, but they have two fundamental issues linked to the rule. The first is lock time. In the previous example, the desired channel spacing was 1 kHz, giving a 1-kHz PFD frequency. Because the output of the PFD contains spikes of energy at the 1-kHz frequency, the low-pass filter between the PFD and the VCO must have a cutoff frequency that is significantly lower than 1 kHz in order to reject this noise. Typically, a good starting point is to set the loop filter bandwidth at 10% of the PFD frequency, so let's assume that we use a 100-Hz filter.

What happens if you reprogram the N divider to switch the PLL to another frequency? The output of the PFD will change, but this analog signal will have to pass through the 100-Hz low-pass filter prior to changing the VCO frequency, which will take time because the filter is filtering out all quick variations. The PLL will lock to the new frequency, but probably some hundreds of milliseconds later. So with an integer-N PLL you have the choice between small frequency steps and fast lock time, not both.

The second issue with integer-N PLLs is a little less straightforward and related to phase noise. Basically, a PLL "multiplies" the PFD frequency by N. As you have seen, in order to have small frequency steps, you need a low PFD frequency, which means you need high values for N. Unfortunately, the phase detector is not a perfect device, so it generates some phase noise, which is unfortunately multiplied by N. Trust me, the PFD noise is increased by 20 log(N) in decibels, so it is far higher if you need to increase the N divider ratio. With an integer-N PLL, small frequency steps will always mean a more noisy output and a longer lock time. Life is difficult.

Fortunately, engineers have developed a new kind of PLL that doesn't have these intrinsic limitations: so-called fractional-N PLLs. The idea is simple. A fractional-N PLL allows integer values for the N divider as well as fractional values. That's a great idea. Through fractional-N PLLs the same channel spacing can be achieved with higher values for the PFD frequency, giving a quicker lock time and far lower noise.

Let's take the previous example and assume that you have a fractional-N chip enabling a 0.01 resolution for N. You can then use a PFD frequency of 100 kHz (compared to 1 kHz with the integer-N PLL). An output frequency of 500 MHz + 1 kHz would be achieved with $N = 5000.01$; the loop filter would be 50 kHz and not 500 Hz (providing a 100× improvement in lock time), and the noise due to the PFD would be reduced by 20 log(100) = 40 dB. This seems magical, but nothing is free. No one has found a way to build a perfect fractional frequency divider on silicon (at megahertz or gigahertz speeds).

The trick used by actual fractional-N chips is to dynamically switch N between two integer values (e.g., 5000 and 5001 in this example) with a time ratio proportional to the designed fractional value. Note that 99% of the time with $N = 5000$ and 1% of the time with $N = 5001$, roughly a 5000.01 divide ratio is achieved. Unfortunately, this introduces some noise on the PFD output, which can generate nasty spurious frequencies on the output spectrum. Moreover, because of implementation restrictions, a fractional-N chip is more complex to program and can't usually generate any arbitrary frequency.

PLL Design

In summary, the engineering rule is quite simple. If you need to generate several frequencies spaced by a constant—a reasonably large frequency step (as compared to the inverse of the wanted lock time)—and if you don't have stringent noise requirements, integer-N PLLs will help. Just set the PFD frequency equal to the desired frequency step size and start with a low-pass loop filter with a cutoff frequency

10 times lower. However, if you need a fine frequency resolution or have stringent phase noise requirements, fractional-N solutions could be a better answer, but be ready to accept some spurious frequencies on the output spectrum and spend a little more time tweaking their setting registers. The best performances with a fractional-N PLL are often achieved with a high PFD frequency because the setting will minimize lock time and PFD noise. But that isn't always the case. As with the integer-N PLLs, the low-pass filter should usually be set at around one-tenth of the PFD frequency.

This first approach will give you a good starting point for your design. However, a more detailed analysis is needed to verify that the design works and to optimize the PLL parameters and loop filter to get the best performance. Fortunately, the main PLL chip suppliers have developed free simulation tools that make this exercise a pleasure. An example is shown in Figure 10.7 with Analog Devices's ADIsimPLL tool suite, which is available on the company's Web site.

A tool such as ADIsimPLL enables you to define the system-level specifications of your desired PLL such as minimum and maximum frequencies, channel spacing, reference frequency, and so on. It then automatically proposes adequate chips (of course from the manufacturer who has offered you the tool) and compatible external VCOs if needed. After you select a loop filter and a couple of other parameters, your PC shows you the schematics for your PLL subsystem and the full simulation results both in time domain (lock time) and frequency domain (phase noise).

I must be honest with you. You may dislike the simulation results because they probably will not be what you have dreamed of. That's life. But you can go back and change parameters, such as PFD frequency or loop filter bandwidth, until you are fully satisfied. The good news is in the next step: Such simulations are accurate, so you shouldn't encounter too many surprises when you switch on your soldering iron, at least if your PCB is well designed. Remember that millivolts of noise on the VCO input can generate visible phase noise or spurious signals on the output.

Silicon Trends

Let's talk silicon for a minute. The first integrated PLL was probably the NE56x series, originally from Signetics, but they are a little obsolete in comparison to today's

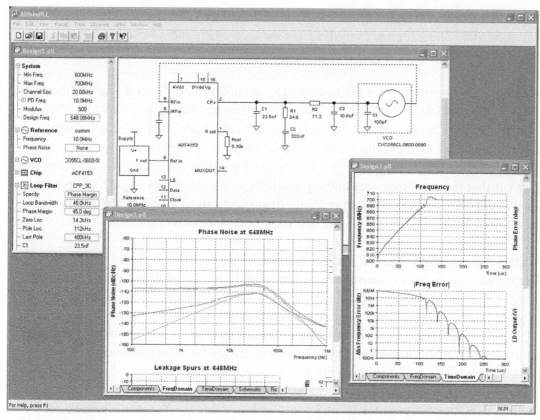

Figure 10.7: This screenshot shows you how pretty PLL design tools such as ADIsimPLL (Analog Devices) can be. Full PLL schematics (*top*) calculated by the tool (*bottom left*), a frequency domain simulation, and (*bottom right*) a time domain simulation.

standards. Nearly all manufacturers have PLLs in their catalog, so I will not give you a list of all the chips on the planet. For an example, go to *http://www.analog.com/pll* to see the PLL product line from Analog Devices. It will give you an idea of the PLL performance you can expect to find on the market, even if you can also find similar products from Texas Instruments, Maxim Integrated Products, and others.

A couple of years ago, most PLL chips were integrating the dividers and PFD, but they were designed to drive an external VCO through an external loop filter. This is still the case for high-performance chips, but solutions are now available that also integrate the

VCO. For example, chips such as the Texas Instruments TRF3761 and the Analog Devices ADF4360 integrate everything but the loop filter, and, of course, the simulation tools of their respective suppliers support the parts.

Another very interesting trend is mixing PLL technology and direct digital synthesis (DDS). What is DDS? Well, I will explain it in the next chapter, but basically it's a digital way to generate variable-frequency signals. PLLs are great in high frequencies, but they have difficulties with small frequency steps. It is exactly the opposite with DDS, so a wedding between the two technologies can give decent results. And the good news is that those solutions are already available in silicon. An example? Chips such as the Analog Devices AD9956 can generate frequencies up to 2.7 GHz with subhertz resolution. They're not for everyday applications, but they can be useful.

At the other extreme, you can find medium-performance PLL chips that are easy to use (requiring nearly no external components) and impressively flexible. For example, assume that you are designing a new multimedia digital device. You have done a system-level design, and you discover that you need four different clock frequencies: 10, 25.55, 8.42, and 5.44 MHz. Are you going to buy four custom crystals? That would be the wrong decision because Cypress Semiconductor has chips such as the CY22393, which includes no less than three full PLLs, their associated VCOs, a 4×4 crosspoint switch, five programmable extra dividers, and file clock output drivers, all in a space-saving TSSOP16 package for $4 or $5 each in low volumes! You don't believe me? Have a look at Figure 10.8, which shows the low-cost Peppermint evaluation kit. You can configure it with the easy-to-use PC-based CyberClocksRT software and a USB cable (see Figure 10.9).

Lastly, Figure 10.10 shows you the actual output waveform, which is not too bad with a single crystal.

It's Your Turn!

As I said in my introduction, PLLs are everywhere. Their immediate use is to generate a fixed-frequency clock or RF carrier based on a reference frequency. Take a look at your latest motherboard. There isn't a high-frequency oscillator. The approximately 3-GHz clock is generated inside the processor itself. However, a PLL's application range is far

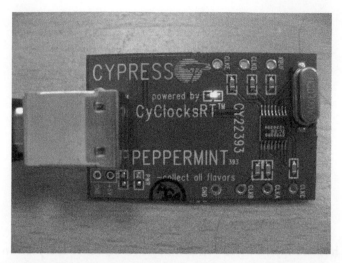

Figure 10.8: Cypress Semiconductor proposed a low-cost evaluation kit for its CY22393 multi-clock generator. All the required components to build a five-output clock generator are visible in the picture. A USB interface chip is on the bottom face.

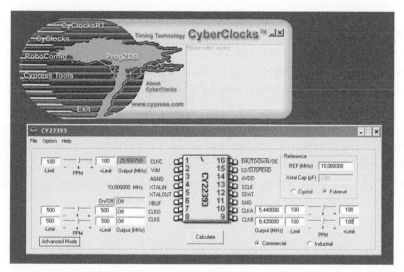

Figure 10.9: The CyberClocksRT software enables you to easily configure the CY22393 chip and download the parameters to the evaluation board through a USB link. Of course, not all frequencies can be generated; however, the tool shows you the frequencies that are not fully exact and the error in ppm.

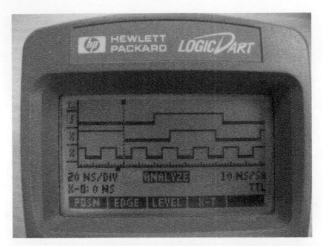

Figure 10.10: I hooked my Hewlett-Packard LogicDart pocket analyzer on three outputs of the CY22393, and, yes, I got the desired 5.44, 8.42, and 25.55 MHz frequencies from top to bottom, respectively.

wider. Digital transmissions (from Ethernet to ADSL to wireless protocols) all transmit their bits of information serially and rely on PLLs to recover the clock from the signal itself. Because the transmitted signal has frequency content at the carrier frequency, a PLL with a locking range around this carrier frequency will quickly lock on it.

By the way, to ensure that the lock will always occur, the data stream must have regular bit transitions even if the data is always 0 or 1. This is why bit-encoding methods, such as Manchester encoding or 8/10 bit encoding, are used everywhere. PLL can also "clean" a noisy clock signal, act as a nearly perfect FM demodulator, and be used in a zillion other applications.

Wrapping Up

Full books can be written on this subject. I can't cover everything in a single chapter, but my hope is that the subject is now demystified for you and that you will not be frightened to think "PLL" for your next design. PLLs are not black magic, even if they are sometimes on the darker side.

CHAPTER 11

Direct Digital Synthesis 101

In Chapter 10, I talked about using a phase-locked loop (PLL) to generate precise and stable frequencies. I also briefly introduced another interesting concept: the direct digital synthesizer (DDS). It is now time to dig into that subject and help you understand how DDS techniques, with their pros and cons, can help you in future projects. So come with me on a journey to DDS World.

DDS Basics

The simplest form of a digital waveform synthesizer is a table look-up generator (see Figure 11.1). Just program a period of the desired waveform in a digital memory (Why not an EPROM for old timers?), connect a binary counter to the address lines of the memory, connect a DAC to the memory data lines, keep the memory in read mode, clock the counter with a fixed-frequency oscillator F_{CLOCK}, and, voila, you've got a waveform on the DAC output. Don't forget to add a low-pass filter to clean the output signal, with, as you know, a cutoff frequency a little less than $F_{CLOCK}/2$ to please Mr. Nyquist.

This design works, but it is not too flexible. If you want to change the output frequency, you need to change the clock frequency, which is not easy to do, especially if you need a fine resolution. The DDS architecture is an improvement on this original design (see Figure 11.2).

Note: This chapter is a corrected and enhanced reprint of the article "The darker side: Direct digital synthesis 101," *Circuit Cellar*, no. 217, August 2008.

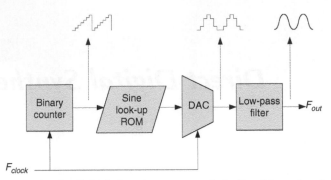

Figure 11.1: The most basic digital signal generator is built with a simple binary counter. Its output sequentially addresses the rows of a memory, which holds the successive points of the output signal. It is then converted to an analog signal and filtered.

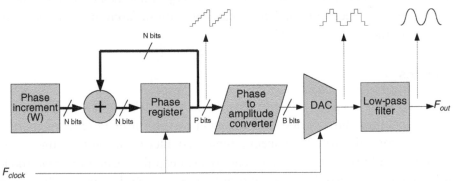

Figure 11.2: The basic architecture of a DDS is a variant of the counter-based digital generator, but it allows a fine frequency resolution, thanks to a phase register and a binary adder. The key point is that the increment is not necessarily a divider of the phase register maximum value.

Rather than add 1 to the table look-up address counter at each clock pulse as in the previous example, a DDS uses an N-bit long phase register and adds a fixed-phase increment (W) at each clock pulse to it. N can be quite high (e.g., 32 or 48 bits), so only the most significant bits of the phase register are used to select a value from the phase-to-amplitude look-up table, which is usually nothing more than a ROM preprogrammed with a sine waveform. Assume that you are using the P most significant bits as an address. Then the output in the look-up table is routed to a DAC. And, of course, the analog signal finally goes through a low-pass filter, which is called a "reconstruction filter." You will understand why in a minute.

How does it work? If the phase increment W is set to 1, you will need 2^N clock pulses to go through all of the values in the look-up table. One sine period will be generated on the F_{OUT} output each 2^N clock pulses, exactly as in the aforementioned counter-based architecture. If W is 2, it will be twice as fast and the output frequency will be doubled. As you know, you need a little more than two samples per period to be able to reconstruct a sine signal, so the maximum value of W is $2^{N-1} - 1$. The formula giving the output frequency based on the phase increment is then

$$F_{OUT} = W \times \frac{F_{CLOCK}}{2^N}$$

Don't be confused. It is not a simple programmable divider because the phase register doesn't loop back to the same value after each generated period. The phase turns by a fixed angle on the so-called phase wheel at each clock pulse. Figure 11.3 may help you understand it.

What makes a DDS a fantastic building block is the numeric examples. Just take a standard, low-performance DDS with a phase register of $N = 32$ bits and a reference clock $F_{CLOCK} = 20$ MHz. Your DDS can then generate any frequency from DC to nearly 10 MHz with a resolution of the following:

$$1 \times \frac{20 \text{ MHz}}{2^{32}} = 0.0046 \text{ Hz}$$

Not bad. In fact, the maximum frequency will be a little lower because of constraints on the low-pass filter, as you will see later.

DDS Flexibility

Another great advantage of a DDS generator is that you can use it for any kind of modulation, while still fully in the digital domain. Refer to Figure 11.4, which shows a slightly enhanced DDS architecture.

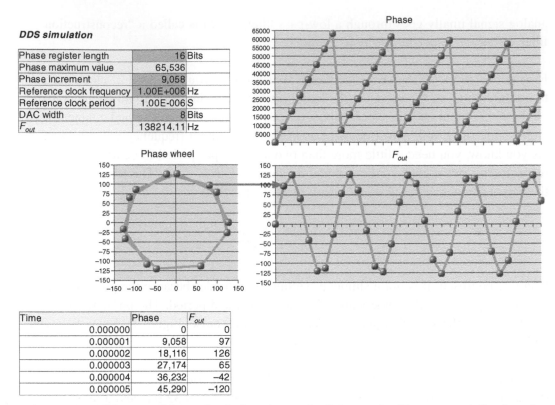

DDS simulation

Phase register length	16	Bits
Phase maximum value	65,536	
Phase increment	9,058	
Reference clock frequency	1.00E+006	Hz
Reference clock period	1.00E-006	S
DAC width	8	Bits
F_{out}	138214.11	Hz

Time	Phase	F_{out}
0.000000	0	0
0.000001	9,058	97
0.000002	18,116	126
0.000003	27,174	65
0.000004	36,232	−42
0.000005	45,290	−120

Figure 11.3: This spreadsheet simulation shows the "phase wheel" concept. A fixed angle is added to the phase register at each clock pulse. Note that no period of the output signal is identical to the previous ones because the phase doesn't go back to the same value after a full turn.

With a DDS, you can easily change the output frequency on the fly, without any delay or phase shift, just by loading a new value in the phase register or by switching between different phase registers for FSK-like transmissions. You can also add a fixed value to the phase register independently of the DDS itself, which is ideal for phase modulation or PSK. You can add a digital multiplier before the DAC to implement software-controlled amplitude modulation.

You can generate waveforms other than sine just by loading a period of your designed signal in the look-up table. But in that case, be careful: You will be drastically limited

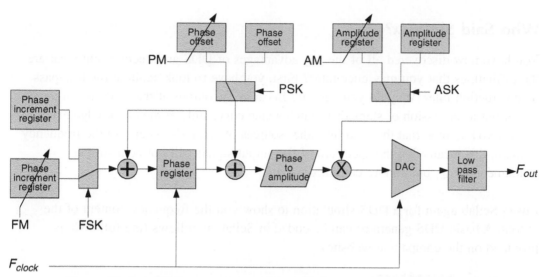

Figure 11.4: A DDS generator can be easily improved to add full digital modulation features, including frequency (FM), phase (PM), or amplitude modulations (AM).

in terms of maximum frequency. Because of the mandatory output low-pass filter, all harmonics above the Nyquist limit of $F_{CLOCK}/2$ will be filtered out. So for non-sine signals, you will be limited to output frequencies low enough to ensure that all harmonics required for a good generation of your signal are significantly below this limit, which usually means frequencies not above $F_{CLOCK}/20$ to $F_{CLOCK}/100$.

As an example, look at the datasheet for a lab-class arbitrary signal generator such as the Agilent Technologies 33220A, which is 50 Msps. It states that the maximum sine frequency = 20 MHz and the maximum triangle frequency = 200 kHz. Now you know why. If you need to generate a square signal, you will not have these limitations because you can generate a sine and add a simple comparator to extract a square signal with the same frequency.

There are a lot of other possibilities, thanks to the digital structure of a DDS, and silicon makers are imaginative in these areas. You will see some examples later on.

Who Said Sin(x)/x?

You have now discovered all of the key advantages of DDS architecture, but what are the difficulties that you may encounter? First, you have to look again at the low-pass reconstruction filter. Why do you need it? Because the output of the DAC is not a sine signal but a succession of steps that match a sine curve only at the clock-edge events, even if you assume that there are no other sources of error elsewhere. In the frequency domain, this means that the spectrum of the output signal will not be a simple fundamental F_{OUT} but a more complex signal.

I used Scilab again for a DDS simulation to show you the frequency content of the output. A basic DDS generator can be coded in Scilab as follows (the full code is provided on the companion website):

```
// Global parameters
phase reg bit count = 16;
clock freq = 1e6;
dac bit count = 8;
simulation steps = 1000;
nyquist bands for spectrum calculation = 6;

//----------------------------
// Phase increment value, which sets the frequency
phase increment = 5169;

// Deducted parameters
phase reg max = 2^phaseregbitcount;
dac max = 2^(dacbitcount-1);
time increment = 1/clock freq;

// steps vector
steps = 0:simulation steps-1;

// Time vector
time = steps*time increment;

// Phase vector, ie phase for each time step
phase = modulo(steps*phase increment, phase reg max);

// Output signal, through sine lookup vector
scaled phase = phase*2*%pi/phase regmax;
fout = int(dac max*sin(scaled phase));
```

```
// output signal with better time resolution through
  interploation
fout_ext = 1:simulationsteps*nyquist bands for spectrum
  calculation;
for i = 1:simulation steps*nyquist bands for spectrum
  calculation,
fout_ext(i) = fout(1+(i-1)/nyquist bands for spectrum
  calculation);
end;

// calculate spectrum
spectrum = fft(fout_ext);
spectrum = 2*spectrum/(simulationsteps*nyquist bands for spectrum
  calculation);
// However as the input signal is real then only the first half
  of the FFT
// is actually useful (the last half is the complex conjugate
  of the first half):
useful spectrum = spectrum(1:$/2);
spectrum amplitudes = abs(useful spectrum);
spectrum amplitudes(1) = dac max; //just to scale the graph

// sin(x)/x reference for display
x = [1:nyquist bands for spectrum calculation*simulationsteps/2];
x = x*%pi/simulation steps;
sinxoverx = x;
for i = 1:length(x), sinxoverx(i) = dac max*abs(sin(x(i))/x(i));
  end;

// Plot signals
subplot(3,2,1); plot(phase(1:50)); xtitle('phase (N=16, P=16,
  B=8, W=5169)');
subplot(3,2,3); plot(fout_ext(1:50*nyquistbandsforspect
  rumcalculation)); xtitle('Fout');
subplot(3,2,5); plot(1:length(spectrumamplitudes),
  spectrumamplitudes,'k',1:length(spectrumamplitudes),
  sinxoverx,'g'); xtitle('spectrum');
```

Load this code under SciLab, click "execute," and you've got Figure 11.5.

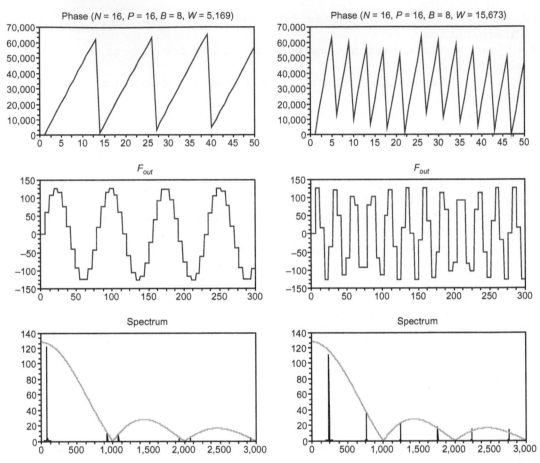

Figure 11.5: This Scilab simulation shows the phase, output signal, and output spectrum of a 16-bit DDS clocked at 1 GHz, with two different tuning words (F_{OUT} = 78 MHz on the left and 239 MHz on the right). The amplitude of the fundamental frequency gets lower when the frequency increases—following a sin(x)/x curve (shown in superposition)—and image frequencies become more powerful and unfortunately closer to the desired frequency.

As you can see, there are image frequencies in the output. You get not only the frequency F_{OUT} but also $F_{CLOCK} - F_{OUT}$ and $F_{CLOCK} + F_{OUT}$, and even $2\,F_{CLOCK} - F_{OUT}$ and $2\,F_{CLOCK} + F_{OUT}$, and more, and you usually need to filter out these unwanted signals through the filter that follows the DAC, namely the reconstruction filter. The respective amplitudes of these image frequencies follow a curve mathematically defined as $\sin(x)/x$, which happens to be the Fourier transform of a single step of width $1/F_{CLOCK}$.

But there is another problem. As shown by the $\sin(x)/x$ shape, when your output frequency increases, the power of the image frequencies also increases. As power needs to be found somewhere, this implies that the power of your desired F_{OUT} signal becomes lower and follows the same $\sin(x)/x$ curve shown in Figure 11.5. This leads to two problems.

One, you need to know (and compensate for if necessary) the reduction of signal amplitude when the frequency gets closer and closer to the Nyquist limit, at which point the theoretical power reduction is 3.92 dB. Two, when you come close to this limit, the first image frequency, which you need to cancel out with the low-pass filter, comes closer to your desired frequency and, worse, at similar amplitude. Because the required low-pass filter is impossible to build, you can't actually generate a signal arbitrarily close to the $F_{CLOCK}/2$ limit (see Figure 11.6). The usual reasonable limit is around 40% of F_{CLOCK} even with sharp filters.

However, nothing prevents you from using one of these image frequencies instead of the fundamental. Just replace the low-pass filter with a bandpass filter and you can use a DDS to generate a frequency higher than the Nyquist limit, far in the ultra-high-frequency area. The amplitude will be lower, but it will work as long as your filter is well designed.

Any Other Problem?

Once you have managed to filter out any image frequencies, will you get a perfectly clean sine signal? You will, but only if you have a perfect DDS with an infinite number of bits and infinite precision everywhere. Unfortunately, you are not that rich. One of your enemies will be DAC resolution. Because the resolution B of the DAC is not so high, there will be a quantization error, which will translate into quantization noise in the output spectrum. Once again, I have a small Scilab simulation with two different

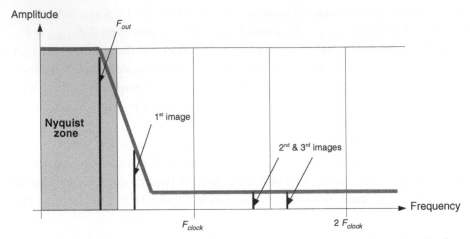

Figure 11.6: The role of the low-pass filter is fundamental. It must keep the fundamental frequency nearly untouched while providing a high attenuation on all image frequencies. That's why straight filters are usually required. That's also why the maximum frequency is usually 40% and not 50% of F_{CLOCK}.

DAC resolutions and using basically the same code as before. The result is shown in Figure 11.7.

The theory says that the power ratio of signal to total quantization noise is $1.76 + 6.02B$ dB, with B as the resolution in bits of the DAC. For example, with an 8-bit DAC, you can expect a 50-dB (i.e., $1.76 + 6.02 \times 8$) signal-to-noise power ratio. But that's just an average. However, there is a trick if quantization noise is a problem.

Because the noise is somehow spread from DC to the Nyquist limit, you can limit it with just a bandpass filter around your frequency of interest. If you reject all frequencies except a 10% passband around F_{OUT}, the quantization noise will be divided by 10. Another solution is oversampling. If you increase F_{CLOCK} without increasing the low-pass filter corner frequency, the quantization noise will also be lower in the filter passband.

DDS has another issue that's often more crucial than quantization errors: phase accumulator truncation. The number of bits in the phase accumulator register is not infinite; nor is the number P of input bits in the look-up table. This will give another

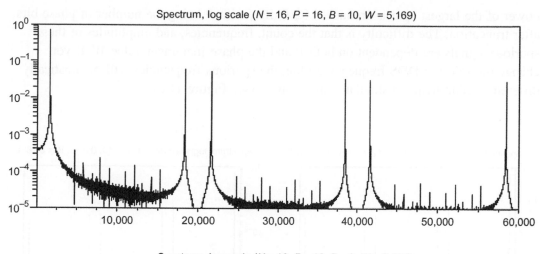

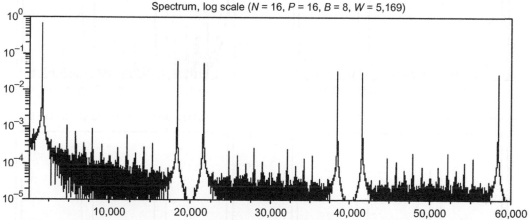

Figure 11.7: This simulation, also done with Scilab, shows the DAC quantization effect. These are simulated output spectra of a 1-GHz 16-bit DDS tuned to provide a 78-MHz output. The top curve is with a 10-bit DAC. The bottom curve is with an 8-bit DAC. The vertical scale is logarithmic.

error on the output. Contrary to DAC quantization, this error will generate not broadband noise but discrete spurious frequencies on the output spectrum. You may think of it as a miniature unwanted DDS generator working on the unused bits and unfortunately added to the output. Once again, the theory helps. It says that the relative power of the largest spur is around −6.02 *P* dBc, with *P* being the number of phase bits after truncation. The difficulty is that the count, frequencies, and amplitudes of these spurious signals are dependent on both *P* and the phase increment value *W*. If you change the selected DDS frequency a little, the spurious frequencies will be drastically different. Scilab helps to simulate this behavior (see Figure 11.8).

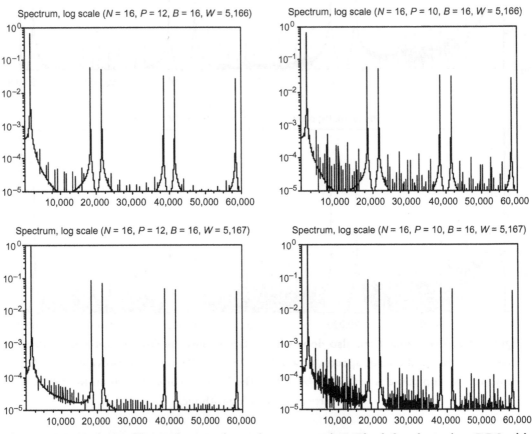

Figure 11.8: Here I'm illustrating spurious phase truncation. The left column is a DDS with 12 bits effectively used as a look-up table address. The right column has only 10 bits. The spurious spectrum is far more numerous and powerful in the latter. Finally, the bottom line shows what happens in the same condition with just a small change in the tuning word value (5167 versus 5166). The spectrum of spurs is different.

Behavior like that shown in the figure makes life for a DDS designer a little more complicated, but it is also a potential friend. If you have some flexibility in the DDS parameters, for example, and if you can get a slightly different F_{CLOCK} or F_{OUT}, then you may find another combination that gives fewer spurious frequencies (or at least fewer spurs in a given frequency band). The good news is that the behavior of the DDS is predictable and some good simulation tools are available from chip manufacturers.

There are other sources of noise in DDS (clock jitter, DAC nonlinearity, clock feed-through, and the like), but image frequencies, DAC quantization, and phase-register truncation are usually the main contributors. However, don't conclude that DDS generates only noisy signals. These problems exist, but a good, well-designed DDS can have signal-to-noise ratios well above 70 dBc, large enough for the vast majority of applications. By the way, you will find two different figures in the specifications: the signal-to-noise ratio and the spur-free dynamic range. They are correlated but not equivalent. The former is the ratio of signal power to the sum of all noises. The latter is the ratio of signal power to the strongest spurious frequency.

Software Implementation

Enough theory. It's time to demonstrate how to build an actual DDS generator. There are some impressive dedicated integrated circuits around. I will examine them later. For now, let's start with a firmware-based implementation.

Programming a DDS in a general-purpose microcontroller or DSP is often an effective solution for signal generation. Imagine that you are using a small microcontroller (e.g., a Microchip Technology PIC16F629A clocked at 20 MHz) and you need to generate a 7117-Hz signal, either sine or square. This is just an example, of course, but real-life applications can include DTMF generation, data rate generation, and similar problems. The first idea that you, a firmware developer, will have is to use the on-chip timer. Just configure a timer to count processor cycles (5 MHz maximum on this PIC variant) and toggle the output per N cycles. Calculate N for the output to be as close as possible to the required 7117 Hz. Here you can have either 7122 Hz (i.e., 5 MHz/702) or 7112 Hz (i.e., 5 MHz/703). That isn't too bad, but it's quite far from the target, and you can't program a "fractional" count on a timer.

This is where DDS helps. Imagine another approach: Configure the on-chip timer for an interrupt at any frequency but significantly above 2×7117 Hz (e.g., 50 kHz). At each interrupt, add a fixed amount W to a 16-bit phase register, convert it to a sine using an 8-bit ROM-based look-up table, and send the value to a DAC. Then filter it with a 10-kHz low-pass filter. Refer to the schematic in Figure 11.9, in which

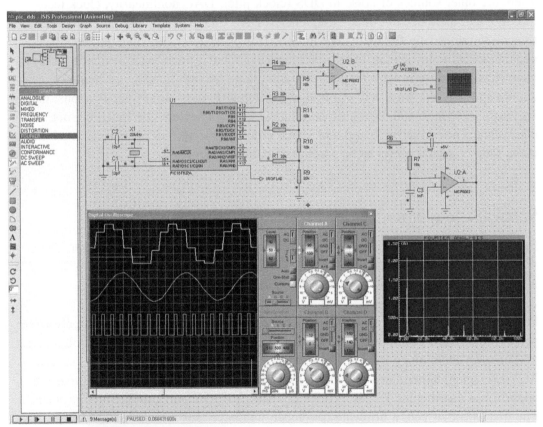

Figure 11.9: This is the Proteus VSM in action. The schematic shows you that I used a PIC microcontroller, a small R/2R 4-bit DAC, and an active filter. Just virtually load the associated firmware in the microcontroller, connect a virtual oscilloscope or spectrum analyzer to the output, click "run," and you have the simulated output on the display. Impressive.

I just used a simple 4-bit passive R-2R network as a DAC and a pair of Microchip MCP6002 op-amps as a buffer and low-pass filter. If you need a square signal, you can simply route the filtered signal back to the comparator available inside the PIC.

At this point, I can't resist telling you about a great simulation tool for mixed-signal designs. Labcenter Electronics's Proteus tool suite includes tools for schematic entry, Spice simulation, and PCB design. It also provides an impressive simulator named Virtual System Modeling (VSM) as an option. With VSM, you can simulate the code running on a microcontroller, as with any firmware simulator, and the electronic circuits, as with any Spice-like simulator, but you can do both simultaneously. Take another look at Figure 11.9. A virtual scope enabled me to verify the DDS signals generated by the PIC and filtered by the MCP6002, all without having to switch on the soldering iron! With the advanced simulation option, Proteus VSM can even calculate the spectrum of the filtered signal, as shown in Figure 11.9. Close to expectations, isn't it?

The associated source code, fully coded in C using the free Hi-Tech Software PICC-Lite compiler, is very short:

```
// Firmware DDS demonstration
// Author   : R.Lacoste for Circuit Cellar
// Target   : 16F627A
// Compiler : Hitech PICC-Lite 9.60
// IDE      : MPLAB 8.10

#include    <htc.h>
#define TIMER_VAL (0xFF-100+13)   // Timer reload each 100 cycles
   (20µs, or 50KHz)

                        // +13 to compensate interrupt latency

unsigned short phase;        // Phase register is 16 bits
unsigned short phase increment;    // Phase increment is also 16 bits

// Sine table, from 0 to 255
const unsigned char SINE_TABLE[]=
```

```
{
128, 131, 134, 137, 140, 143, 146, 149, 152, 156, 159, 162,
  165, 168, 171, 174,
176, 179, 182, 185, 188, 191, 193, 196, 199, 201, 204, 206,
  209, 211, 213, 216,
218, 220, 222, 224, 226, 228, 230, 232, 234, 236, 237, 239,
  240, 242, 243, 245,
246, 247, 248, 249, 250, 251, 252, 252, 253, 254, 254, 255,
  255, 255, 255, 255,
255, 255, 255, 255, 255, 255, 254, 254, 253, 252, 252, 251,
  250, 249, 248, 247,
246, 245, 243, 242, 240, 239, 237, 236, 234, 232, 230, 228,
  226, 224, 222, 220,
218, 216, 213, 211, 209, 206, 204, 201, 199, 196, 193, 191,
  188, 185, 182, 179,
176, 174, 171, 168, 165, 162, 159, 156, 152, 149, 146, 143,
  140, 137, 134, 131,
127, 124, 121, 118, 115, 112, 109, 106, 103, 99, 96, 93, 90,
  87, 84, 81,
79, 76, 73, 70, 67, 64, 62, 59, 56, 54, 51, 49, 46, 44, 42, 39,
37, 35, 33, 31, 29, 27, 25, 23, 21, 19, 18, 16, 15, 13, 12, 10,
9, 8, 7, 6, 5, 4, 3, 3, 2, 1, 1, 0, 0, 0, 0, 0,
0, 0, 0, 0, 0, 0, 1, 1, 2, 3, 3, 4, 5, 6, 7, 8,
9, 10, 12, 13, 15, 16, 18, 19, 21, 23, 25, 27, 29, 31, 33, 35,
37, 39, 42, 44, 46, 49, 51, 54, 56, 59, 62, 64, 67, 70, 73, 76,
79, 81, 84, 87, 90, 93, 96, 99, 103, 106, 109, 112, 115, 118,
  121, 124
};

#define PHASEINCREM
// timer 0 interrupt routine
void interrupt timer0_isr(void)
{
unsigned char look up index;

TMR0=TIMER_VAL;                         // Reload timer
PORTA=0b00001;              // Signal interrupt processing on RA0
phase+ = phase increment;        // Increment the phase register
look up index = (unsigned char)(phase >> 8); // get the top 8 bits
PORTB = SINE_TABLE[look up index];   // and convert it to a sine
PORTA = 0b00000;
T0IF = 0;                               // Clear interrupt flag
}
```

```
// Main program
main()
{
    // initialize DDS
    phase = 0;
    phase increment = 9328; // fout = 50 KHz ×
  phasevincrement/65536 = 7116,69 Hz

// initialize timer 0
OPTION = 0b1000;     // timer mode, no prescaler
T0CS = 0;            // select internal clock
TMR0 = TIMER_VAL;    // Initialize timer
T0IE = 1;            // enable timer interrupt
GIE = 1;             // enable global interrupts
TRISB  = 0;          // port B is output
TRISA = 0b11110;     // RA0 is output
PORTB = 0;           // Port B start at 0

for(;;)
    continue;        // Do nothing, all is done in the interrupt
}
// That's all, folks.
```

You have built an actual DDS, and you can generate any frequency calculated as

$$F = W \times \frac{50 \text{ kHz}}{65,536}$$

and thus any frequency from 0.76 Hz to close to 20 kHz with a frequency step of 0.76 Hz! For example, just choose $W = 9328$ to get a frequency of 7116.69 Hz. That's far closer to the 7117-Hz target, isn't it? The magical trick comes from the fact that a DDS allows drastically finer frequency steps because the phase increment is not necessarily a sub-multiple of the period.

Silicon Solutions

Even if software-based DDS is possible, there are plenty of impressive silicon versions, particularly from Analog Devices. One example is an easy-to-use chip such as the AD9833, a $4 (qty. 1,000-2008) low-power chip fitted in a 10-pin 3 mm × 3 mm

MSOP package that enables you to generate frequencies up to nearly 12.5 MHz with 0.1-Hz resolution. The chip is driven by a standard serial peripheral interface (SPI) port, which enables you to connect it to any microcontroller. Since energy consumption is such an important consideration, note that this chip consumes no more than 5.5 mA at 3 V, which is impressive for a 25-MHz chip. Technically speaking, the chip has a 28-bit phase register, a 12-bit look-up table, and a 10-bit DAC, which provides around 60 dB of signal-to-noise ratio. Now you can easily understand such a datasheet.

Let's focus on a current, top-of-the-line DDS chip. The Analog Devices AD9910 depicted in Figure 11.10 is nearly 10 times more expensive than the AD9833. It costs around $35 (qty. 1,000-2008). But what a piece of silicon! First, its clock can be as high as 1000 MHz, providing a useful output range of up to 400 MHz. Providing a 1-GHz clock may be difficult, but the guys at Analog Devices had the good idea to include an on-chip PLL to allow more reasonable external clock sources. The chip's 32-bit phase accumulator provides sub-hertz resolution, and it is equipped with a high-speed 14-bit DAC, enabling a spur-free dynamic range of up to 65–70 dBc and still around −55 dBc at 400 MHz. But that was for the DDS core alone, and this chip has plenty of other blocks.

First, it has an auxiliary DAC to define the full-range amplitude without compromising the quantization noise. It can also automatically compensate for the $\sin(x)/x$ amplitude roll-off I discussed earlier, with a digital filter that has an inverse $\sin(x)/x$ response placed between the look-up table and the DAC. You can program eight different settings for frequency, phase, and amplitude, and then switch among them in nanoseconds via three external pins. If necessary, the chip can also automatically manage linear frequency, phase, or amplitude sweeps. In addition, it has a built-in 1024×32 RAM that enables you to predefine custom frequency/phase/amplitude profiles and execute them at high speeds, which is perfect for generating complex modulated waveforms.

What else? Oh yes, it can be synchronized with other chips if its features are not enough for your application. This is managed through an SPI, but it also has a high-speed parallel bus for time-critical applications. Okay, it is not a low-power chip (800 mW, 1.8 V, and 3.3 V), and you will have to solder its 100 pins and read its 64-page datasheet if you decide to use it, but that's the price you pay for such a list of features.

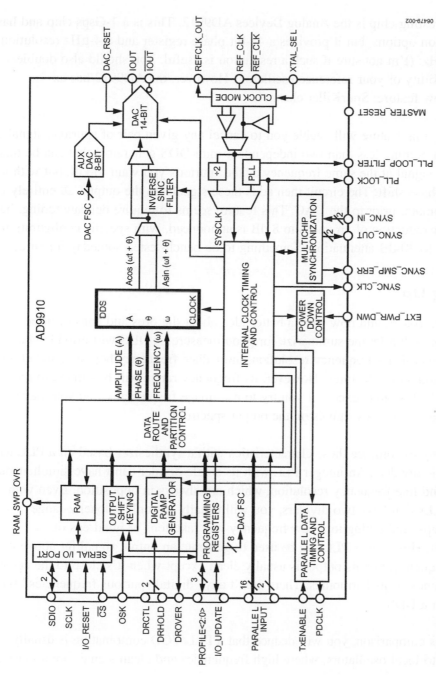

Figure 11.10: This is the internal architecture of the AD9910 high-end DDS chip (courtesy of Analog Devices). As you see, the DDS core is just a small part of the chip. It is surrounded by a zillion advanced high-speed digital modulation and control blocks, as well as a 14-bit Gsps DAC and a reference clock PLL multiplier.

Another interesting chip is the Analog Devices AD9912. This is a 1-Gsps chip and has less modulation options, but it provides a 48-bit phase register and a 4-µHz resolution up to 400 MHz. (I'm not sure if such a resolution is useful. You should also double-check the stability of your reference oscillator.) However, the AD9912 has an interesting new feature: SpurKiller channels.

Theoretically, this feature will enable you to cancel any given pair of spurious signals. It is based on nothing less than two independent mini-DDS generators that can be tuned to generate a signal at the same frequency as the spurious you want to kill, but with a 180-degree phase shift. The circuit then adds these signals on the output, all entirely in the digital domain, prior to the DAC. This feature seems to require delicate tuning, but a typical spur reduction of around 6 to 8 dB is announced, with specific configurations providing up to 30-dB attenuation. Something to be bench-tested someday for sure!

Wrapping Up

Here you are. You should now have a better idea about the pros and cons of direct digital synthesis, but let me summarize for good measure. A DDS will provide you with a marvellous sub-hertz frequency resolution, immediate frequency hopping, and efficient full-digital modulation features. However, its frequency range will be limited to around 40% of the clock source, except if you try to use image frequencies, and you may suffer from some nasty spurious signals on the output spectrum.

It is interesting to compare these characteristics with a synthesizer based on a PLL with programmable dividers. An integer PLL with its single divider can't have simultaneous fast tuning and fine frequency resolution, which are always in opposition. Even with fractional PLLs that have two dividers, you will usually get only kilohertz-range frequency steps, and tuning to a new frequency will take tens or hundreds of microseconds. However, a PLL can be used to generate an output signal far above its reference frequency, and its output is usually clean except when it's close to the center output frequency or its harmonics. There aren't any "digital spurious frequencies" such as those with a DDS.

Based on this comparison, you will deduce that a PLL/VCO combination is usually more suited to local oscillators, where high frequencies and clean signals are a must. A

DDS finds its key application as a modulation source where agility is most important. However, DDS chips have gotten cleaner and cleaner over the years, and nobody would have imagined seeing a 1-GHz DDS chip for tens of dollars a couple of years ago. For the best of both worlds, there are chips with both PLL and DDS cores. So stay tuned—things can change quickly! Now it's your turn. You should be ready to put a DDS in your next design, either as a piece of silicon or as some lines of firmware. DDS is no longer on the darker side for you!

Part 5
Communications

CHAPTER 12

Open Your Eyes! A Primer on High-Speed Signal Transmission

Communication, even simple wired communication, is another interesting dark side subject. You are used to connecting your RJ-45 plugs to the nearest router and assuming they will work without any problems—and they usually do. You understand that wireless networks such as Wi-Fi can be a little trickier, yet even reputable electrical engineers think that wired communications are, well, just wires. But have you ever thought twice about the black magic that actually enables you to send and receive 100 Mbps, 1 Gbps, or even 10 Gbps on these small twisted-copper pairs? Similarly, do you know how to use a low-cost FR4 PCB and connectors costing less than $1 to build your PC motherboard with its 2.525-Gbps PCI-Express 2.0 busses?

In this chapter, I will present basic techniques for high-speed signal transmission on low-bandwidth cables. I will cover concepts ranging from eye diagrams to pre-emphasis and equalization. You may find this information useful for your next high-speed design.

High-Speed Issues

Suppose you have to design a transmitter and a receiver to send a binary data stream through a roll of twisted-pair wire. As an engineer, you want to ensure that your design will work, so you first measure the wire's actual transmission channel. A simple technique involves injecting a sine signal with a sweeping generator at one end of the

Note: This chapter is a reprint of the article "The darker side: High-speed signal transmission—from eye diagrams to pre-emphasis and equalization," *Circuit Cellar*, no. 227, June 2009.

cable and then connecting a receiver, oscilloscope, or power meter at the other end. Since you take care of impedance matching, you can easily measure the cable's 3-dB cutoff frequency, which is the frequency at which the received signal decreases to 50% of its original power.

Imagine that you are lucky enough to find a 100-MHz cutoff frequency, which is already good for a standard copper pair. You want to transmit binary signals and not just pure sines, so you need to transmit frequencies significantly higher than the bit rate. That is, you must cope with a couple of harmonics to have well-shaped bits, which means more or less rectangular waveforms. So you must use a bit rate reasonably lower than 100 MHz. Ten times lower than the cutoff frequency, giving 100 MHz/10 = 10 Mbps, may be a reasonable guess. Therefore, you will go back to the marketing department and announce that the system will be 10 Mbps.

You know what will happen next, right? The answer will be something like "Are you joking? We need 1 Gbps." And then your life becomes far more difficult as a result of an unpleasant phenomenon: intersymbol interference (ISI). Figure 12.1 shows where ISI comes from.

Basically, a cable used closer to or higher than its cutoff frequency behaves like a highly damping RC network. The rising and falling edges of any transmitted bit are reshaped with exponentially decaying slopes, with a time constant that can be significantly longer than the bit duration. That's a problem. But the most critical consequence is that the received voltage for a given transmitted bit will depend not only on this specific bit but also on the previously transmitted bits. This will inevitably generate erroneous bits on the output because a binary receiver is basically nothing more than a comparator to a given threshold.

For example, if you send alternating 0s and 1s, the output signal can be distorted but will still cross the comparator threshold each time and give a solid transmission. However, if you send a long string of 1s, the data line voltage will reach its maximum even if the RC time constant is long. Then, if you send only one 0 followed by a new set of 1s, the line voltage will decrease a little but not enough to be identified as a 0 by the receiver. This "history" phenomenon is the basic form of ISI, and it can actually make error-free reception nearly impossible.

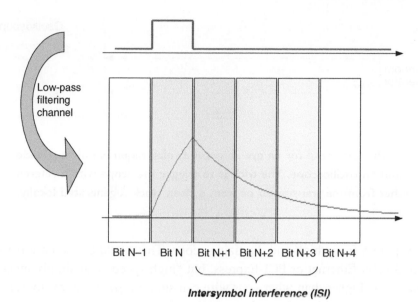

Bit N–1 Bit N Bit N+1 Bit N+2 Bit N+3 Bit N+4

Intersymbol interference (ISI)

Figure 12.1: ISI appears because of the transmitting cable's relatively long time constant. The line voltage is perturbed several bits later than the original bit. This spreading makes reception far more difficult.

Do You See This Eye?

The intersymbol issue is not new. It was a problem with telegraphic lines right from the start. Telecom workers needed a way to visualize and evaluate a signal's ISI level, so they invented the eye diagram. This powerful, intuitive method needs just two pieces of equipment. First, a pseudo-random binary sequence generator must be connected to the transmitter's input to ensure that all possible ISI scenarios are sent through the channel. Such a pseudo-random generator is available as part of many telecom test sets, but you can build one with a shift register and a couple of XOR gates. If you search the Internet for "Linear Feedback Shift Register," you'll find plenty of schematics, but more on that later.

You also need a simple single-channel oscilloscope. It must have an external trigger input, and its bandwidth must be significantly higher than the signal's bit rate. You shouldn't think that "high-speed techniques" are only for multi-Gbps systems that will

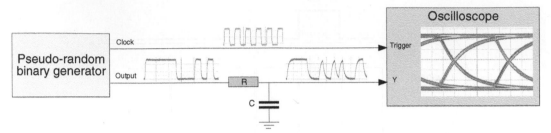

Figure 12.2: The test setup for an eye-diagram display requires only a pseudo-random generator and an oscilloscope. The trick is to trigger the scope with a reference clock, either from the transmitter or from a clean clock regenerated locally.

inevitably require a fancy high-end digitizing scope. That could be the case if you're working on Gigabit Ethernet or PCI-Express, but "high speed" can simply mean your required bit rate is higher than what your wire can support (as in my example). So these techniques are also valuable with medium-rate transmissions (e.g., a few megabits per second), at least if you have to use a long or poor transmission line.

The eye-diagram test setup requires you to connect the scope input to the signal to be measured, and trigger it with the original clock (see Figure 12.2). If you can't—for example, if both ends of the cable aren't in the same room—you can regenerate the clock locally through a phase-locked loop. The oscilloscope's horizontal time setting is set to twice the bit duration, and the trigger delay is adjusted to have a single bit centered on the display. What do you see on the oscilloscope's display? Basically a superposition of bit transition waveforms, which will include every possible bit sequence thanks to the pseudo-random generator. I used my Lecroy Waverunner 6050 oscilloscope. It was driven by the pseudorandom generator section of an old Schlumberger 7703B 2-Mbps bit-error tester through a simple adjustable RC filter. The results are shown in Figure 12.3.

If the RC time constant is low—meaning that the transmission channel is good in comparison to the bit rate—then the shape of the data signal is rectangular and the eye diagram shows only a simple image made with the superposition of the four possible bit transitions in the pseudo-random stream (0 to 1, 1 to 0, 0 to 0, 1 to 1). What happens if you send this signal to a comparator-based receiver? Ideally, the threshold will be centered vertically and you will get an ideal bit stream as an output, without any time jitter. The

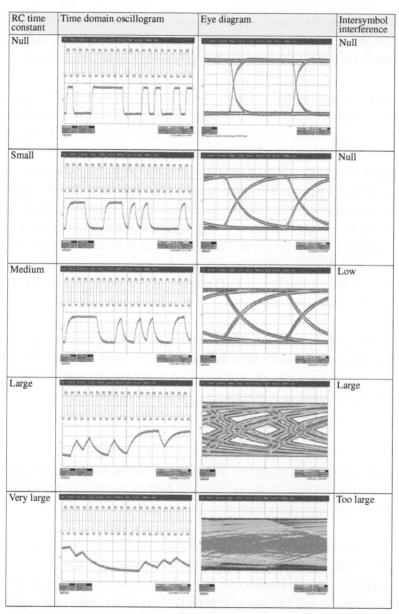

RC time constant	Time domain oscillogram	Eye diagram	Intersymbol interference
Null			Null
Small			Null
Medium			Low
Large			Large
Very large			Too large

Figure 12.3: These are actual signal waveforms corresponding to an eye diagram when the line's RC constant is increased step by step. The ISI percentage is the ratio, on the center horizontal line, between the width of the transition region (where plenty of lines are superimposed) and the bit period. It ranges from 0 at the top to 100% at the bottom.

ISI is 0%. As long as the RC time constant stays low in comparison to the bit period, the image looks the same. The center hole in the eye diagram, the center of the "eye," is said to be fully open, which means that bit recovery by the receiver will be easy.

However, if you increase the RC constant, the eye diagram starts to be more complex (the third case in Figure 12.3). The rising and falling fronts are so damped that the signal can't reach their steady level before the arrival of the next bit, which means that there will be some ISI. The ISI percentage can be evaluated as the ratio, on the horizontal line, between the width of the transition region (where plenty of lines are superposed) and the bit period. It is there around 5%. The more the RC time constant is increased, the more shut the eye with the fourth test case, a comparator, even with a perfectly centered threshold, will show a very high level of time jitter on its output, which means a higher risk of bit error. There is around 50% ISI. At a given point the eye is fully shut, which means that a comparator-based detector will no longer be able to reliably detect if the bit is a 0 or a 1. The transmission no longer works, ISI is 100%, and the eye is fully shut.

The eye diagram enables you to quickly evaluate the ISI level and the link's reliability. Using it, I can also give you more data than just the ISI level. For example, the channel noise margin is simply proportional to the eye's height. The allowed level of timing errors, including clock jitter, is proportional to the eye's width.

Pre-emphasis

Assume that you are in a bad situation: The eye diagram is fully closed when you have a mandatory wire and bit rate. How can you still build a reliable transmission system without compromising the requirements?

Think twice. You are in trouble because the channel (your wire) has a low-pass-shaped transfer function, with a cutoff frequency that is too low. The answer seems obvious. You need to amplify the high frequencies to get back to an overall flat frequency response end to end. As you can see in Figure 12.4, you have two main options. You can amplify the high frequencies on the transmitter side with a technique called "pre-emphasis," or you can do it at the receiver side with a technique called "equalization." Many systems use both techniques together for optimum results.

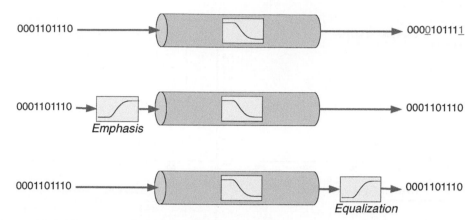

Figure 12.4: The correction of the channel transfer function can be on the transmitter side (pre-emphasis), on the receiver side (equalization), or on both.

Let's focus first on the transmitter side with pre-emphasis. The ISI comes from the shape of the pulses, which are too slow to rise or fall. This means that there isn't enough high-frequency energy during the bit transitions. The most usual and clever trick for signal emphasis uses only a limited amount of electronics. Whenever there is a bit transition (i.e., whenever a transmitted bit is different from the previous one), you can simply boost its voltage. This works by using not just two but four voltage levels: strong high, strong low, weak high, and weak low. For example, if the line is 0 and you have to transmit a couple of 1s, your circuit first sends a strong 1 for the first bit (e.g., 5 V) and then a lower voltage for the next 1s (weak 1, 4 V, or so). Similarly, when a couple of 0s have to be sent after a 1, the first will be a strong 0 (e.g., 0 V) and the next bits will be weak 0s (approximately 1 V). This way, a little more energy is given during each transition because of the strong bits.

As a consequence, the rising is quicker and the long-term damping is reduced, and it can be nearly zeroed with a proper emphasis ratio. Let's simulate emphasis under Scilab for a respectively short and long sequence of logic 1, with a channel simulated as a simple first-order low-pass filter. The result is shown in Figure 12.5 (Scilab source code is provided on the companion website). If you look at the frequency domain, you see that such a pre-emphasis scheme increases the signal's high-frequency components, which is exactly what you are looking for.

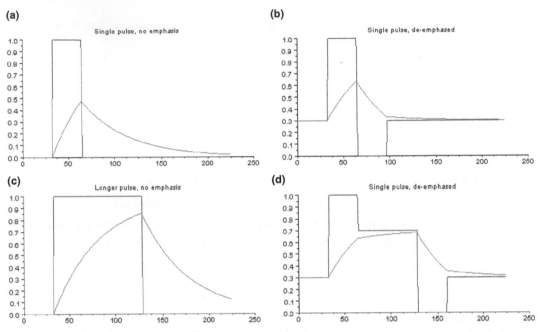

Figure 12.5: This is a Scilab simulation of pre-emphasis. The rectangular-shaped curves are the transmitted waveforms. The attenuated curves are the received waveforms. The left column is without pre-emphasis, respectively with a single bit (a) or a sequence of 1s (c). It is clearly visible on the right (b and d) that the pre-emphasis enables drastic reduction of the ISI.

Just to be exhaustive, note that two variants of pre-emphasis can exist. In the first, the voltages of the strong bits are actually increased in comparison to the standard logic voltages. This has an important drawback because you will need higher-than-normal and negative voltages on your transmitter. The most usual way to get them is to use standard logic voltages for the strong bits and reduced voltage swing for the weak ones, with a fixed and preset ratio that's optimized depending on the transmission line and bit rate (e.g., by looking at an eye diagram), an option usually called de-emphasis.

If you remember Chapter 7 on finite impulse response (FIR) filters, you understand that such a pre-emphasis is actually equivalent to a two-tap FIR filter (the output is a linear combination of the current and previous samples). More evolved emphasizing techniques are possible with more evolved FIR filters. The idea is, in fact, to add to the waveform to be transmitted a delayed and reversed copy of the erroneous channel response. But I won't cover that here.

Can a simple pre-emphasis solve the problem in my RC filter example? Yes. Let's simulate it under Scilab:

```
//––––––––––––––––––
// Parameters
//––––––––––––––––––

// Number of simulation periods for pulse response
periods pulse = 7;

// Number of simulation periods for random signal
periods = 100;

// Number of points per period
points per period = 32;

// RC filter coefficient
rc = 0.02;

// Emphasis coefficient
emphascoeff = 0.3;

//––––––––––––––––––
// Filtering function
//––––––––––––––––––

function out = rcfilt(in)
  v = in(1);
  for k = 1:length(in)
    i = (in(k)-v)*rc;
    v = v+i;
    out(k) = v;
    end;
end function

//––––––––––––––––––
// Pseudo-random signal generation
//––––––––––––––––––
```

```
// Binary pseudo-random signal
binsignal = grand(periods,1,'uin',0,1);
binsignal(1:2) = 0;
input signal = 1:periods*points per period;
for(i = 1:periods*points per period)
  input signal(i) = binsignal((i-1)/points per period+1);
end;

//————————————————————
// Generate the de-emphasized signal
//————————————————————

emphased = 1:periods*points per period;
current = 0;
for i=0:periods-1
  if (i = 0) then
    next = emphascoeff+(1-2*emphascoeff)*binsignal(1);
  elseif ((binsignal(i) == 0) & (binsignal(i+1)==0)) then
    next=emphascoeff;
  elseif ((binsignal(i)==1) & (binsignal(i+1)==1)) then
    next=1-emphascoeff;
  elseif ((binsignal(i)==0) & (binsignal(i+1)==1)) then next=1;
  elseif ((binsignal(i)==1) & (binsignal(i+1)==0)) then next=0;
  end;
  for j=i*pointsperperiod+1:i*points per period+points per period
    emphased(j)=next;
  end;
end;

//————————————————————
// Generate output signal through channel simulation
//————————————————————

// Low pass filtered channel
outputnotemphased=1:periods*pointsperperiod;
outputemphased=1:periods*pointsperperiod;
outputnotemphasedsignal=rcfilt(inputsignal);
outputemphasedsignal=rcfilt(emphased);
```

Just add some extra code to display the result as an eye diagram, click "run," and you get Figure 12.6 (full source code is provided on the companion website).

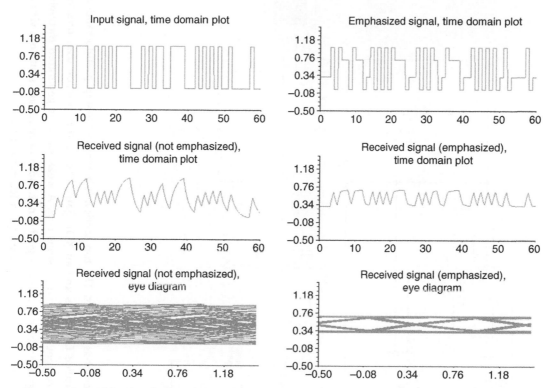

Figure 12.6: This simulation shows the impressive effect of pre-emphasis at least with a transmission channel modeled as an RC network: without emphasis (*left*), with emphasis (*right*). The eye diagram looks better, doesn't it? However, keep in mind that a real-world situation can be a little more complex than this simulation.

This simulation in Figure 12.6 shows that when using a pre-emphasis transmitter, the received signal level is clearly lower, because we are reducing the level of the weak bits, but it has a far lower ISI, which means a far better bit error rate. You may wonder how to actually build a pre-emphasis circuit in real life. I will describe existing dedicated chips later, but you can do it yourself, at least if the signal speed stays reasonable.

Figure 12.7 shows a possible solution that uses old 74xx series chips. As I drew the first version of this schematic with the Proteus CAD suite, I clicked the Simulate icon and got what you see in Figure 12.8. Nice, isn't it?

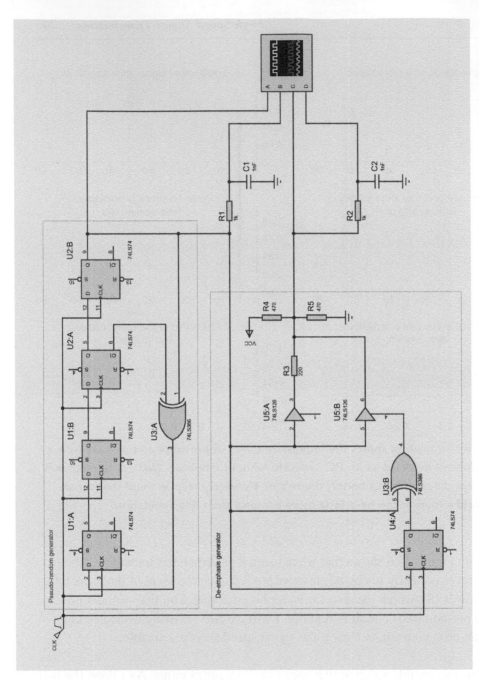

Figure 12.7: This is a discrete pre-emphasis circuit. The top section is a simple 16-step pseudo-random sequence generator made with four so-called D latches and an XOR gate. The bottom portion uses one more D gate and an XOR gate to compare the current bit to the previous one. A weak signal is generated through R3 if both bits are the same. The XOR gate enables a strong signal if not. Both outputs are sent to a virtual scope for simulation (directly and through a RC network).

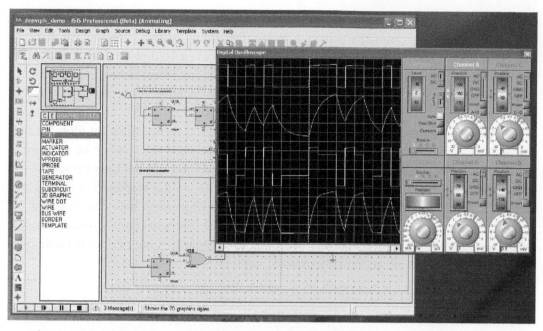

Figure 12.8: This is the result of Proteus VSM simulation of the schematic provided in Figure 12.7. I have not built the circuit, but I am pretty sure the actual measurements would be close.

Equalization

Equalization is an option that involves building some kind of selective high-frequency booster on the receiver side. It is usually less of a magical solution than pre-emphasis because a high-pass filter has the disadvantage of amplifying any high-frequency noise on the line. That is why equalization is typically used as a complement to pre-emphasis. Pre-emphasis does 80% of the job, then equalization finishes it.

On the hardware side, an equalizer is often simply a digital FIR filter called a feed-forward equalizer. More complex nonlinear algorithms exist, such as the distributed feedback equalizer, but a FIR is usually enough.

Equalization provides a great bonus: A receiver actually sees the degraded received signal, whereas a transmitter can't, at least without a dedicated feedback information channel. It is then theoretically possible for the receiver to adjust the equalization filter to compensate for the channel's behavior and implement a "perfect" channel. In that case, if you know the problem, you can fix it. This can even be done if the channel's behavior changes over time, as is the case on a wireless link. This technique, called adaptive equalization, is used in many modern wireless systems.

Without using a complex adaptive filter, how can you build an equalization filter for my example RC simulation? See Figure 12.9. As shown, you need to calculate or measure the channel's frequency response (including pre-emphasis if any). It's then easy to deduce the amplification factor over frequency required for a flat response. A simple Fourier transform gives you the coefficients required to implement the corresponding FIR filter. (Once again, please refer to Chapter 7 on FIR filters to refresh your memory.) I did it in Scilab for you (see Figure 12.10). The eye diagram is not perfect because of the filters, but at least the signal is improved.

Some Silicon

Of course, silicon suppliers have ready-made chips for implementing pre-emphasis or equalization. Few solutions exist for relatively low-speed data links, but notable exceptions are Maxim Integrated Products' MAX3291 and MAX3292. These chips look like classic RS-485/422 transceivers, except that they include pre-emphasis circuitry to allow higher bit rates on long cables. The MAX3291 is programmable for preset data rates of 5 and 10 Mbps, while the MAX3292 is configurable up to 10 Mbps through an external resistor. The corresponding Maxim application note (AN643) states that these chips enable you to double the bit rate in comparison to standard RS-485 drivers.

You can also find dedicated high-speed equalizer chips, but they are scarce. Examples are National Semiconductor's EQ50F100n and DS64EV100. The latter is an equalization filter that improves data links up to 10 Gbps on cables or on a standard FR4 PCB through a choice of eight preset filter settings. The National Semiconductor DS40MB200 is a dual 4-Gbps buffer with both programmable pre-emphasis and a fixed equalization filter. It's available from Digi-Key if you want to play with it. However, the majority of pre-emphasis/equalizer applications use FPGAs, not dedicated chips. This makes sense because these techniques are mainly useful for high-speed systems, and FPGAs are the de facto choice for such projects.

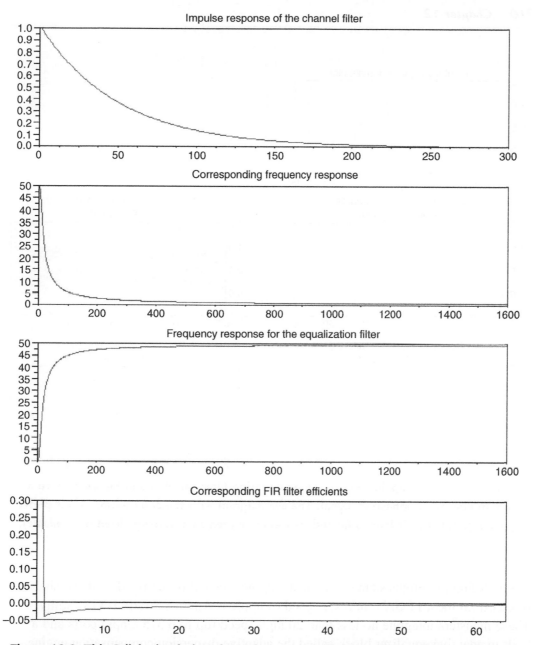

Figure 12.9: This Scilab simulation shows how an equalization filter could be generated. The top plot is the time domain impulse response of our reference RC channel. Take its FFT and you get the second plot. The horizontal axis is the frequency. This is the transfer function of the transmission channel. Negate it to get an ideal compensation network, in the third plot. You can then take its inverse FFT, which will directly give you the required impulse response of the equalization filter, the bottom plot, which is nothing less than the coefficients of the equalizing FIR filter.

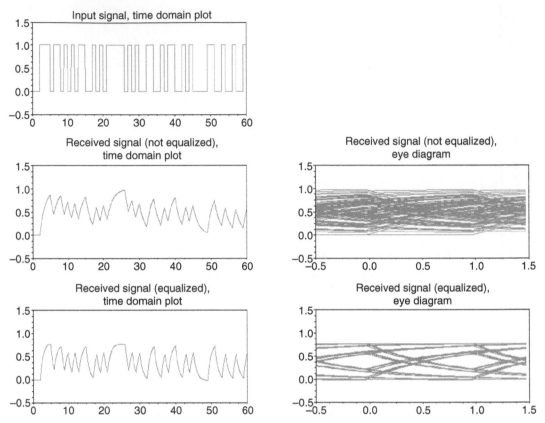

Figure 12.10: An application of the equalization filter calculated in Figure 12.9 to a random non-pre-emphasized signal. The eye diagram with the equalization is not perfect (*bottom*), but the ISI level is far reduced as compared to a non-equalized one (*middle*).

You will find pre-emphasizing and equalizing intellectual property (IP) off-the-shelf blocks from most FPGA vendors. A good example is Altera's Stratix II GX family. These monsters integrate no less than 20 high-speed (up to 6.375 Gbps) ports and a built-in adaptive equalizer block called the adaptive dispersion compensation engine, which uses an impressive mixed-signal approach. In a nutshell, a capacitor-based booster amplifies bit transitions on each input, with a boosting effect controlled by a low-speed DAC. Your application can tune the DAC for a minimal bit error rate.

A last word: Equalization can be analog, too. Nothing prevents you from designing an analog high-pass filter that can more or less exactly compensate for the frequency behavior of a given cable. It can be difficult to do, but it is possible. For example, Tyco Electronics proposed a line of high-speed interconnect cables with a built-in equalization PCB through Tyco's HSSDC high-speed serial data connector family. These cables are specified for no less than a 2-Gbps transmission up to 30 m away on a copper cable. Try to do that without equalization.

Wrapping Up

The techniques that have been covered in this chapter are used in nearly all high-speed communication systems. Roughly speaking, a simple "strong/weak bit" pre-emphasis technique can boost the high frequencies. It can be a good fit for the less complex situations you encounter. Equalization, and adaptive equalization in particular, enables you to fine-tune a channel for optimal performance. It's a must whenever a given channel must be used at its best. This is quite often nowadays.

You won't need these techniques for every design. If you need them to implement standard high-speed transmission protocols (e.g., Gigabit Ethernet or PCI-Express), you'll probably get them as part of a chipset, so it should be more or less transparent for you. But if you need to work on an exotic data transmission project and can't achieve more than 70% of the target's performance, knowing the basics of pre-emphasis and equalization may enable you to come up with the missing 30%. Don't forget that "high speed" just means quicker than the maximum the transmission channel can natively accept, which may be quite slow. And don't forget the marvelous eye-diagram investigation method.

I hope you found this review of high-speed transmission techniques interesting and useful. Be sure to keep one eye in the time domain and the other in the frequency domain. If you do, pre-emphasis and equalization should no longer be on the darker side.

CHAPTER 13

Digital Modulations Demystified

Digital transmissions aren't new. When I hooked my first 300-bps modem to my Apple II back in 1979, I spent hours just actually listening to the bits coming out of the phone and looking at the blinking LEDs, impressed to discover a new way of exchanging software and data without having to physically meet and swap floppy disks! Now I use roughly the same phone line but at a speed of 12 Mbps, thanks to my ADSL triple-play box. Similarly, on the wireless side we are now able to send more than 100 Mbps through a very-low-cost Wi-Fi link, a significant improvement over the first Telex-on-Radio data transmission systems and their 45.5 bps back in the 1930s.

Do you think these amazing improvements are simply a consequence of Moore's law and faster processors? My Apple II and its 1-MHz 6502 processor would have some issues managing a 100-Mbps stream—but this is only half of the story. The main driving factor is probably the impressive progress made by mathematicians and engineers on the digital modulation side: We can now use the same transmission channels far more efficiently.

Have you already encountered acronyms such as GMSK, OQPSK, QAM, and OFDM? Do you know what they actually mean? If not, then just come with me because in this chapter, I will present the modulations probably used in your latest wireless or wireline transmission gadget.

Note: This chapter is a reprint of the article "The darker side: High-speed signal transmission—Digital modulations demystified," *Circuit Cellar*, no. 233, December 2009.

© Elsevier Inc.
DOI: 10.1016/C2009-0-20196-6

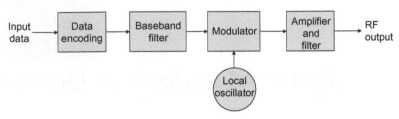

Figure 13.1: In nearly all data transmission systems, the message is encoded, filtered, and then used to modulate a fixed-frequency carrier before amplification and transmission.

Modulation?

Let's consider a basic wireless monodirectional data transmitter. You have a message, which is a finite binary string of zeros and ones, and you want to send it over the air. You will need to build a four-step design as illustrated in Figure 13.1.

First, you will need to encode your data stream. Usually, you will add some preamble and synchronization bytes to help the receiver detect the start of a frame, and a checksum to flag erroneous frames. You will also encode the data itself in a format that is adequate for transmission. This could be simply sending a high level for ones and a low level for zeros, a basic technique called non-return to zero (NRZ). However, NRZ is problematic if you have long strings of zeros or ones because the receiver may lose its clock.

You can also use more robust self-clocking schemes such as Manchester encoding, in which bit values are coded on rising or falling transitions (i.e., a one is coded as "10" and a zero is coded as "01"), but at the expense of a reduced bit rate. You can also use more optimized but more complex encoding such as 8B10B (8 bits coded on 10 bits) or add forward error correction and data-spreading techniques, but that would need a chapter by itself.

Following the data-encoding phase, the signal, which is still made of zeros and ones, is usually low-pass-filtered (more on that later). It is finally used to modulate an RF carrier frequency before transmission. In this chapter I will just focus on this modulation step because there are plenty of methods to send zeros and ones.

OOK?

The most basic modulation method is called On–Off Keying (OOK). Just shut off the RF carrier if there is a zero to transmit, and send a full-power carrier if there is a one, and you have an OOK modulator. This is of course a form of amplitude modulation (AM), and it is used in many low-cost devices such as garage door openers. Like any AM system, an OOK modulator suffers from having a high susceptibility to noise. Another difficulty is that OOK modulators can't be used for high bit rates because they have a comparatively wide frequency spectrum. I wrote a small Scilab script to show you the frequency spectrum of a single OOK-modulated pulse:

```
// Generate a carrier
Fcarrier = 1000000;
Dt = 1/(fcarrier*5);
npoints = 128;
t = (0:npoints-1)*dt;
cw = sin(2*%pi*fcarrier*t);

// Plot it with its FFT
subplot(3,2,1); plot(cw); xtitle('Carrier');
spectrum = abs(fft(cw)); subplot(3,2,2); plot(spectrumc(1:$/2));

// Generates a pulse
Pulse = zeros(1:npoints);
pulse(16:47)=1;

// Plot it with its FFT
subplot(3,2,3); plot(pulse); xtitle('Pulse');
spectrump = abs(fft(pulse)); subplot(3,2,4);
  plot(spectrump(1:$/2));

// Generates an ask carrier
Ask = pulse.*cw;

// Plot it with its FFT
subplot(3,2,5); plot(ask); xtitle('ASK');
spectrum = abs(fft(ask)); subplot(3,2,6); plot(spectruma(1:$/2));
```

The simulation result in Figure 13.2 shows that the frequency spectrum on an OOK pulse includes, of course, the carrier frequency and plenty of other contributors that are regularly spaced above and below the carrier. Why? Look again at Figure 13.2.

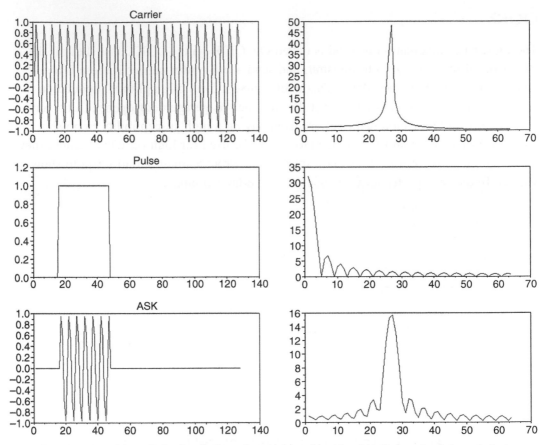

Figure 13.2: This Scilab simulation shows time domain signals on the left and their frequency spectrum on the right. The spectrum of a rectangular pulse is a sin(x)/x shape; the spectrum of an OOK-modulated pulse has the same shape but centered at the carrier frequency.

An OOK signal is in fact the multiplication of the carrier and a one-bit-long rectangular window. Let's switch to the frequency domain.

The frequency spectrum of the carrier is theoretically a single narrow bump. However, if you remember Chapter 8, which is about CIC filters, you may remember that the frequency spectrum of a rectangular window is a curve mathematically defined as

$\sin(x)/x$. It has a main lobe centered at 0 Hz but it also has an infinity of side lobes of decreasing amplitude. The first side lobe is 13 dB below the main lobe, which is indeed quite high. The frequency spacing of the lobes is the inverse of the bit duration (so the higher the bit rate, the wider the spectrum). Finally, mathematicians say that the spectrum of the product of two signals (here the carrier and the rectangular window) is the convolution of their individual spectrums. Convolution may be a difficult concept, but in this case it is simply the $\sin(x)/x$ spectrum of the rectangular window shifted to be centered at the carrier frequency, as shown in Figure 13.2.

That was OOK. Binary amplitude-shift keying (2-ASK) is a variant of OOK where the RF power is not fully null for the transmission of zeros. It can, for example, be switched between 100% and 10% of the full power and allows the probability of errors to be limited in the case of the presence of interferers, but at the expense of a more complex circuit. ASK can also be used with more than two power levels. For example, a 4-ASK modulation uses four different RF powers (e.g., 10%, 40%, 70%, and 100%) to transmit two bits at a time, respectively 00, 01, 10, or 11. This allows doubling of the bit rate as two bits are transmitted at once, but at the risk of significantly more transmission errors.

Baseband Filtering

The issue in RF is usually that you can't use as many spectrums as you want, except maybe if you are working on some military projects. Unfortunately, a modulation such as OOK has a frequency spectrum that goes very far from the carrier because of the $\sin(x)/x$ roll-off. What can you do to use less bandwidth? You can add a filter, of course. The first solution could be to use a narrow bandpass filter on the RF output, precisely centered at the carrier frequency and suppressing all modulation products more than a few kilohertz away from the carrier. This is actually a solution used in some devices with surface acoustic wave or quartz filters, but this is not easy if the product does not have a fixed frequency. The other solution is to filter the signal before the modulator, that is, to filter the baseband zeros and ones as shown in Figure 13.1. Remember that the $\sin(x)/x$ roll-off is due to the window defining each modulated bit. If this rectangular window is replaced by a smoother shape, the spectrum will be cleaner.

What is the ideal filter? It's a filter that allows having, after the modulator, a spectrum constrained to a given frequency band around the carrier and strictly null elsewhere, that is, something like a rectangular window once again but this time in the frequency domain. And what is the time domain impulse response of such a filter? You know it: $\sin(x)/x$, again thanks to the symmetry of the Fourier transform. The spectrum of a rectangular pulse is $\sin(x)/x$, so the spectrum of a $\sin(x)/x$ pulse is a rectangular pulse. Constructing such a filter is difficult, but a good approximation can be achieved if we truncate it after one or two side lobes. Let's simulate it:

```
// Generate a carrier
fcarrier = 1000000;
dt = 1/(fcarrier*5);
npoints = 128;
t = (0:npoints-1)*dt;
cw = sin(2*%pi*fcarrier*t);

// Generates a pulse
pulse = zeros(1:npoints);
pulse(20:20+31) = 1.001;

// Generates an ask carrier
ask = pulse.*cw;

// Plot it with its FFT
subplot(3,2,1); plot(ask); plot(pulse,'r'); xtitle('ASK');
spectrum = abs(fft(ask)); subplot(3,2,2); plot(spectruma(1:$/2));
  xtitle('ASK spectrum');

// generates a raised cosine mask
// its just a cut version of the fft of a pulse…
pulserc = zeros(1:npoints);
pulserc(1:4) = 1/128; pulserc($-3:$) = 1/128;
fftpulserc = fft(pulserc);
lengthrc = 64;
rcfilter = zeros(1:npoints);
for i = 1:lengthrc/2,
  rcfilter(lengthrc/2-i+1) = fftpulserc($-i+1);
  rcfilter(lengthrc/2+i) = fftpulserc(i);
end;

// Plot it with its FFT
subplot(3,2,3); plot(rcfilter); xtitle('RC filter impulse
  response');
spectrum = abs(fft(rcfilter)); subplot(3,2,4);
  plot(spectrumr(1:$/2)); xtitle('RC filter frequency response');
```

```
// Filter the baseband signal
pulsefiltered = convol(pulse, rcfilter);
pulsefiltered = pulsefiltered(1:npoints);

// Regenerate the filtered OOk signal
// Generates an ask carrier
askfiltered = pulsefiltered.*cw;

// Plot it with its FFT
subplot(3,2,5); plot(askfiltered); plot(pulsefiltered,'r');
  xtitle('ASK filtered baseband');
spectrumaf = abs(fft(askfiltered)); subplot(3,2,6);
  plot(spectrumaf(1:$/2)); xtitle('ASK filtered spectrum');
```

The result is Figure 13.3, which shows you drastic improvement in the frequency spectrum of an OOK-modulated pulse when the rectangular window is replaced by such a filter. By the way, this filter is a raised cosine filter, and you will find it easily in the literature. A variant, the root-raised cosine filter, is simply the square root of the former and is used to split such a filter 50% on the transmitter side and 50% on the receiver side, but the behavior is identical. Gaussian filters are also used, but basically any low-pass filter will help.

I've presented baseband filtering in the case of OOK, but the same technique can be used for every other modulation. I will show you an example in a minute.

FSK and Its Variants

Frequency modulation is more resistant than amplitude modulation when noise is added to the signal. As a consequence, binary frequency shift keying (2-FSK) is more robust than 2-ASK or OOK. The idea is just to switch between two closely spaced carrier frequencies $Fc - dF/2$ and $Fc + dF/2$ depending on the bit to be transmitted. Here Fc is the center frequency and dF the modulation width.

What happens on the frequency spectrum? Imagine that you transmit in 2-FSK a single zero followed by a single one. The zero is equivalent to a rectangular pulse modulating a carrier at $Fc - dT$, so on a spectrum analyzer you get the same $\sin(x)/x$-shaped spectrum as a single OOK pulse but centered at $Fc - dF/2$. You get the same for the bit at level 1, but centered at $Fc + dF/2$. The full spectrum of the FSK signal will be the

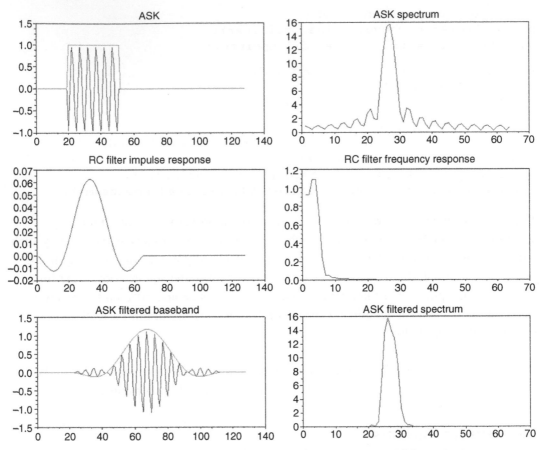

Figure 13.3: As compared to a simple OOK pulse (*top*), the addition of a low-pass baseband filter, here a raised cosine filter (*middle*), drastically limits the frequency width of the modulated pulse (*bottom*).

sum of both shapes, as illustrated in Figure 13.4 (the source code is supplied on the companion website).

The figure shows you another important point. To improve the sensitivity of the receiver, you want to limit the interference between the transmissions of the zeros and ones. Remember Chapter 12 about emphasis and equalization where I presented intersymbol interference? The same problems exist here, but with FSK there

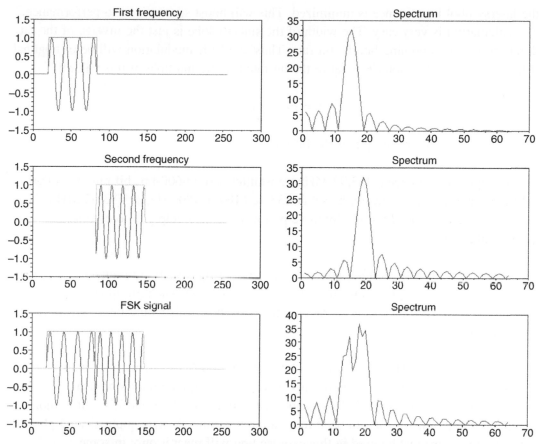

Figure 13.4: The spectrum of an FSK signal is the addition of the spectrums of two OOK-like signals, one centered on *F − dF*/2 and the other on *F + dF*/2. Note that the frequency difference is selected to position the peak of one of the two signals exactly at a null of the other one. This provides orthogonality and improves performance.

is a specific condition that will drastically minimize the problem. Look again at Figure 13.4.

If the separation *dF* between the two frequencies is equal to the exact width of the sin(*x*)/*x* lobe, then the peak of the "zero" spectrum will fall in a null point of the "one" spectrum and reciprocally. The modulation is then called an orthogonal modulation and the intersymbol interference is minimized. This will boost sensitivity and performance. The calculation is very easy: The width of the $\sin(x)/x$ lobe is just the inverse of the bit duration, which is nothing but the bit rate. Thus, the FSK modulation will be orthogonal if the frequency deviation dF is set to the bit rate (or any multiple of this value):

$$F = Fc +/- dF/2$$

where dF = bit rate or multiple of the bit rate.

For example, if you have a 433.92-MHz transmitter and a 9600-bps bit rate, then the 2-FSK frequencies must ideally be set at 433.92 MHz ± 4800 Hz, or 433.92 MHz ± 9600 bps, and so on. This will improve performance and help you to comply with regulations.

Of course, as with ASK nothing prevents you from using only two frequencies in FSK. For example, you can group the signal bits 4 per 4 and code each group as a frequency from a group of 16 frequencies to transmit them at once. This would be a 16-FSK modulation.

A last word on FSK: There is another solution for minimizing intersymbol interference. If you set the frequency deviation to only half the bit rate, then the theoretical interference is in fact also null. This is not visible in Figure 13.3 and it's a little more complex to explain, so you will need to trust me this time. A more sophisticated phase-sensitive receiver must be used in this case, so you will meet it only in some transceivers. Anyway, this specific and very optimized modulation is called minimal frequency shift keying (MSK). By the way, MSK with a Gaussian baseband filter gives GMSK. This is the modulation used in all GSM networks.

I know that you like actual measurements to complement the simulations, so I configured my Agilent E4432B signal generator in MSK mode, using the built-in random signal generator as a modulation source. I then simply connected its output to an Agilent E4406A vectorial spectrum analyzer (I know, I'm lucky.). The result is Figure 13.5, and you will be happy to see that it is very close to the simulation. I then switched on a Gaussian baseband filter and got Figure 13.6. Look at how the spectrum is cleaner.

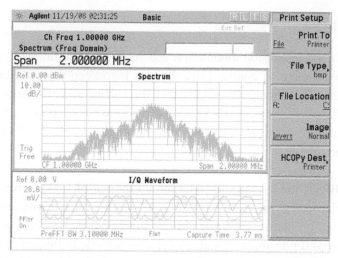

Figure 13.5: This is the actual spectrum of an MSK-modulated 1-GHz carrier, as generated by an Agilent E4432B. It is very close to the 2-FSK simulations shown in Figure 13.4. The bottom plot shows the corresponding *I* and *Q* demodulated waveforms (more on that later). You can see that they are sines with a relative phase of +90° or −90° depending on the bit transmitted.

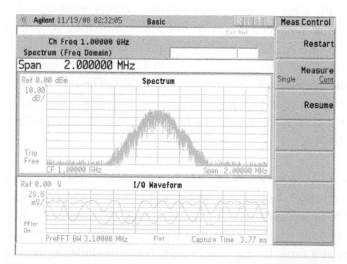

Figure 13.6: The same MSK signal but with a Gaussian baseband filter gives GMSK. The spectral width is much reduced compared with that in Figure 13.5.

Phase Modulation

We have already used amplitude modulation and frequency modulation. What else can we try? Phase modulation, of course. Here the idea is to keep the amplitude and frequency constant while changing the phase of the carrier to distinguish zeros and ones. A binary phase shift keying (BPSK) modulation uses two phases, 0° and 180°, to send zeros and ones, respectively. A signal inverter driven by the bit flow is enough to implement the modulator.

Theoretically, a BPSK modulation allows a more efficient phase-coherent receiver than 2-FSK to be implemented, providing a 3-dB gain in sensitivity. However, phase modulation has two problems. First, the abrupt changes in phase cause a very wide spectrum, so baseband filtering is mandatory. The downside is that such a filter implies that the signal envelope is no longer constant, causing difficulties with amplifier stage, the RF power which can't be a perfectly linear device.

The second issue is more fundamental. On the receiver side, there is no way of knowing the absolute phase of a signal if there is no reference. There are only two solutions to this issue and both are used. In the first approach the protocol must include some specific training sequence to tell the receiver what the reference phase is, and the receiver must then keep it locally. For example, if long sequences of zeros (carrier at phase 0°) are used as a training sequence, then the receiver can lock on this sequence thanks to a local phase-locked loop (PLL) circuit, and can later use this reference to check the phase of the successive data bits. The other solution is to code the information only on relative phase changes and not on absolute phase changes, which is somewhat similar to Manchester encoding. This is called differential phase shift keying (DPSK).

PSK is heavily used because it has another key advantage: It is very easy to use more than two levels without enlarging the spectrum, as in FSK, and without increasing the noise sensitivity too much as in ASK. For example, QPSK uses four phases (0°, 90°, 180°, and 270°) to code two bits at a time, and 8-PSK uses eight phases shifted by 45° to code three bits at a time. By the way, 8-PSK is the modulation used in EDGE systems, allowing a bit rate four times higher than basic GSM. Now you know why: 8-PSK allows transmission of three bits at a time compared with one bit for GMSK, so

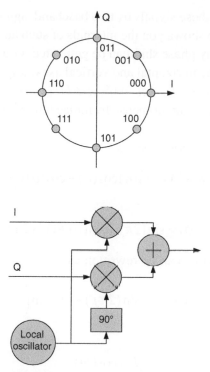

Figure 13.7: (a) An 8-PSK modulation uses eight different phases to encode three bits at a time, here with a Grey code convention. **(b)** An *IQ* modulator is based on two multipliers, each driven by a local oscillator, either in phase or in quadrature. Both signals are then added. This allows generation of any phase shift from 0° to 360° and any amplitude with the proper values for *I* and *Q*.

there is a direct $3 \times$ speed improvement. The remaining 25% improvement is made thanks to other protocol optimizations.

A convenient way to depict a phase modulation is to plot the different states on a polar phase diagram as shown in Figure 13.7a. This is in fact more than a convenient diagram because it actually illustrates the way phase modulators are usually implemented. Rather than trying to shift the carrier frequency by a variable amount, which is technically challenging, PSK systems use a so-called *IQ* modulator architecture. The idea is to use only two versions of the carrier frequency, one in phase and one in quadrature (i.e., shifted by 90°), to multiply each of these signals by two baseband signals, called *I* and *Q*, and to add the results. Figure 13.7b shows you the internals of such an IQ modulator. With the proper values for *I* and *Q*, any phase shift can be generated. Graphically, you just have to read the *I* and *Q* values on the horizontal and vertical axes, respectively. For example *I* = 1 and *Q* = 0 gives 0°, *I* = 0 and *Q* = 1 gives −90°, and *I* = *Q* = 0.707 gives 45°, and so on. Mathematically, the following trigonometric formulae will explain how it works.

One of the basic trigonometric identities is

$$\sin(a+b) = \sin(a)\cos(b) + \cos(a)\sin(b)$$

so

$$\sin(2\pi f + \phi) = \sin(2\pi f)\cos(\phi) + \cos(2\pi f)\sin(\phi)$$

Since cos(a) = sin(a + π/2) this can be rewritten as

$$\sin(2\pi f + \phi) = I \times \sin(2\pi f) + Q \times \sin\left(2\pi f + \frac{\pi}{2}\right)$$

with

$$I = \cos(\Phi)$$

and

$$Q = \sin(\Phi)$$

You recognize the two carriers, in phase and in quadrature, multiplied by the *I* and *Q* values and summed together. By the way, the same circuitry can also be used on the receiver side, as an *IQ* mixer (look at Figure 13.7a from right to left). Such an *IQ* mixer allows an RF signal to be down-converted into two components, *I* and *Q*, without any image issue as with a standard mixer, but that would take us too far from our subject.

Let's simulate an 8-PSK modulation in Scilab using an *IQ* modulator approach:

```
points per period = 16;
bit count=12;
carrier periods per bit = 2;
n points = points per period*carrier periods per bit*bit count;
t = (0:npoints-1);
```

```
// generate I and Q carriers
cw_i=1.001*cos(2*%pi*t/points per period);
cw_q=1.001*sin(2*%pi*t/points per period);

// Generate random words sequence, words from 0 to 3
//bit sequence = int(rand(1,bitcount)*4)+0.001;
Bit sequence=[0 1 2 3 2 0 1 3 2 1 0 3];

// Oversample this signal
Bit sequence_os = bit sequence.*.ones(1,pointsperperiod*carrier
  periods per bit);

// Extract i and q signs
sign_i = 2*int(bit sequence_os/2)-1;
sign_q=2*modulo(bit sequence_os,2)-1;

// Apply sign to each carrier
i_signal = cw_i.*sign_i;
q_signal = cw_q.*sign_q;

// And lastly sum then to get the proper angle
Signal = i_signal+q_signal;

// Plot them all
subplot(4,1,1); plot(bit sequence_os,'b'); xtitle('Coding words');
subplot(4,1,2); plot(sign_i,'b'); plot(i_signal,'r'); xtitle('I
  channel');
subplot(4,1,3); plot(sign_q,'b'); plot(q_signal,'r'); xtitle('Q
  channel');
subplot(4,1,4); plot(signal,'r'); xtitle('Resulting QPSK
  signal');
```

Figure 13.8 shows you the result. QPSK is used in Wi-Fi in its 802.11b 11-Mbps variant as well as in UMTS.

A commonly used variant of QPSK is called offset quadrature phase shift keying (OQPSK). In QPSK, there are four phase states, so *I* and *Q* each have a binary value (+1 or −1). The idea with OQPSK is to limit the phase modifications by changing only *I* or *Q* at a time. Physically, the *Q* signal is shifted half a bit in duration from the *I* signal, and the rest remains identical. Look at Figure 13.9, which shows OQPSK with the same bit stream as in Figure 13.8. OQPSK is used, in particular, for CDMA and for satellite communications.

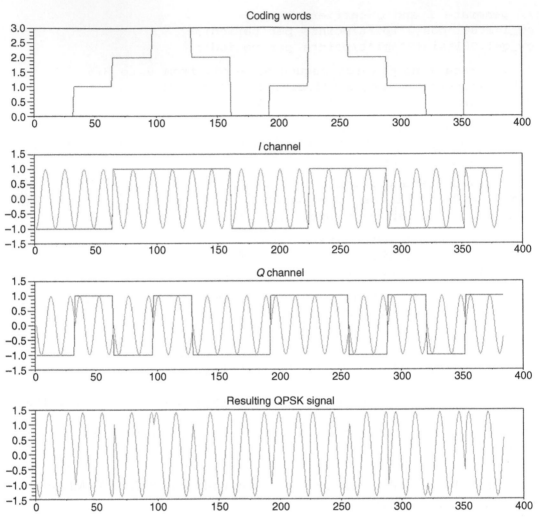

Figure 13.8: This is an example of QPSK modulation. The top shows the bit symbols to be transmitted in each time slot, from *a* to 3. The two middle plots show the *I* and *Q* signals, respectively, and the corresponding output of the multiplier. Lastly the bottom plot shows the resulting modulated signal.

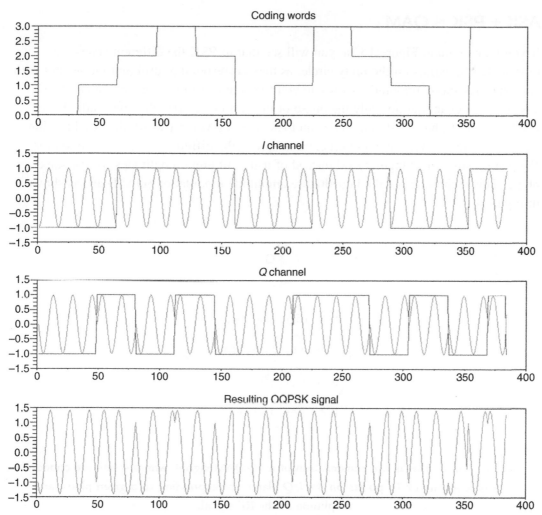

Figure 13.9: OQPSK is a variant of QPSK where the *Q* channel is shifted half a bit on the right as compared to the I channel. Comparing this figure with Figure 13.8, you can see that the phase changes are a little less abrupt.

ASK + PSK = QAM

If you look again at Figure 13.7a, you will see that in PSK the different states are represented by points on the unity circle, as they correspond to different phases but with constant maximum amplitude. How can you transmit even more bits per symbol? The answer is by changing not only the phase of the carrier but also the amplitude. Each combination of phase and amplitude can then code a given bit word, thereby boosting the bit rate. In reality, it is more efficient to spread the different words equally on the *I/Q* phase using a rectangular grid instead of using different amplitudes for the same phase. This technique is called quadrature amplitude modulation (QAM) and is shown in Figure 13.10.

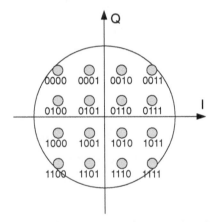

Figure 13.10: This is the constellation of a 16-QAM signal, where 4 bits are coded at a time in one of 16 points on the *I/Q* plane, corresponding to a given phase and amplitude of the RF signal.

Let's simulate a 16 QAM: This is very simple based on our previous 8-PSK example:

```
points per period = 16;
bit count = 12;
carrier periods per bit=2;
n points = points per period*carrier periods per bit*bit count;
t = (0:npoints−1);
```

```
// generate I and Q carriers
cw_i = 1.001*cos(2*%pi*t/points per period);
cw_q = 1.001*sin(2*%pi*t/points per period);

// Generate random words sequence, words from 0 to 3
Bit sequence = int(rand(1,bitcount)*16)+0.001;

// Oversample this signal
bitsequence_os = bitsequence.*.ones(1,pointsperperiod*
  carrierperiodsperbit);

// Extract i and q values
val_i = int(bit sequence_os/4)-1.5;
val_q = modulo(bitsequence_os,4)-1.5;

// Apply sign to each carrier
i_signal = cw_i.*val_i;
q_signal = cw_q.*val_q;

// And lastly sum then to get the proper angle
Signal = i_signal+q_signal;

// Plot them all
subplot(4,1,1); plot(bitsequence_os,'b'); xtitle('Coding words');
subplot(4,1,2); plot(val_i,'b'); plot(i_signal,'r'); xtitle('I
  channel');
subplot(4,1,3); plot(val_q,'b'); plot(q_signal,'r'); xtitle('Q
  channel');
subplot(4,1,4); plot(signal,'r'); xtitle('Resulting QAM16
  signal');
```

Look at Figure 13.11 for the result.

The good news is that the same *IQ* modulator presented in the previous section can be used for QAM. You just have to use more complex combinations of *I* and *Q* signals.

QAM is used heavily in applications requiring a high bit rate in a narrow channel. For example, 16, 32, or even 256 QAM is implemented in many microwave links as well as in digital video standards ranging from DVB-T to DVB-C. Quite impressive. That's because in 256 QAM a full byte is transmitted at once with the selection of one pair of *I/Q* values from a set of 256. Of course, such modulations are more than sensitive to interference and must rely on heavy error correction systems for proper operation.

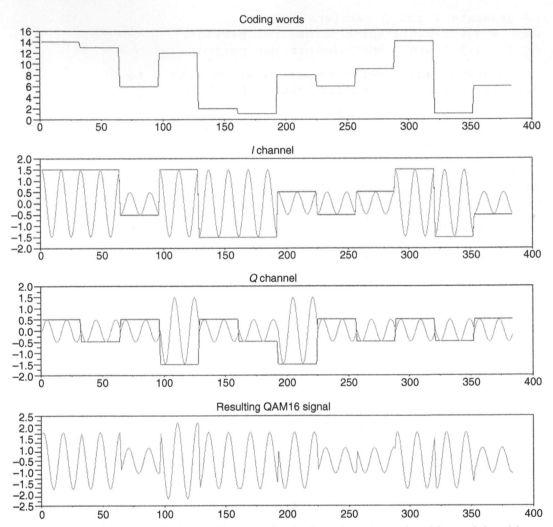

Figure 13.11: A simulation of a 16 QAM shows that the output signal is modulated in both phase and amplitude, giving headaches to power amplifier designers.

From FSK to OFDM

I explained how intersymbol interference can be minimized in FSK just by selecting a frequency deviation equal to a multiple of the bit rate. This configuration allows the peak of one of the two frequencies to be placed in a null of the secondary lobes of the second frequency, providing a so-called orthogonal modulation. The same idea is used in the latest and greatest modulation system, orthogonal frequency division multiplexing (OFDM). There are only two differences. First, OFDM uses not two but hundreds of regularly spaced frequencies. Second, each frequency is used not as a simple switched continuous wave, as in FSK, but as a full transmission channel using any of the previously described modulations, for example, PSK or QAM!

You can imagine that the overall bit rate can be enormous. That's why OFDM is used in ADSL and HomePlug modem systems, Wi-Fi 802.11 n, DAB radios, DVB-H and DVB-T digital videos, WiMAX, WiMedia, and so on. Just as an example, let's see how ADSL2+, now the dominant system used in Europe for triple-play Internet access, works. In ADSL2+, the phone line is used from 0 to 2.2 MHz. This frequency band is split into 512 sub-bands, each of which is 4.3125 kHz wide. For each frequency, a modulation is selected automatically, depending on the performances of the channel in order to transmit from 1 to 15 bits per sub-channel and per time slot. (Think of it as a sophisticated QAM.) Therefore, the maximum bit rate of ADSL2+ is 512×4.3125 kHz $\times 15$ bits, or around 33 Mbps. Not bad on a plain phone line, even if this translates to around 20 Mbps in real life.

Wrapping Up

Digital modulations are not always easy to understand because of the heavy math that sustains them. However, I hope that I was able to present the basic mechanisms in a readable form. And I hope that these techniques are no longer on the darker side for you so you will have one more set of very powerful tools in your pocket!

CHAPTER 14

Antenna Basics

Any RF communication system needs a pair of antennas. What is an antenna? It is simply a device that transforms an electric wave, coming from a generator with a given impedance, into a radiated electromagnetic wave that will propagate in free space or will reciprocally transform an electromagnetic wave into an electric wave to a receiver. Because it is difficult to transmit power over the air, an antenna designer wants reasonable power efficiency in the transmission. To do this, you need to take care of two things: the electric and the electromagnetic sides of the transmission:

- On the electric side, you need good impedance matching between the transmitter or receiver and the antenna or, more exactly, between the transmitter and the combination of the antenna and free space. Reread Chapter 1 if you are not convinced that improper impedance matching will definitely result in inefficient power transfer.

- On the electromagnetic side, because the antenna transforms the electric energy into a radiated field, you need it to radiate its power in the direction that you want. That's the radiation pattern of the antenna.

Impedance Matching—Again

Let's first cover matching and assume that the transmitter or receiver has a standard 50-ohm impedance. Any antenna is fundamentally a resonant system, and its equivalent circuit can be approximated by a serial RLC network, at least around resonance.

Note: This chapter is a reworked version of the column "The darker side: Antenna basics," *Circuit Cellar*, no. 211, February 2008.

Because you read Chapter 1, you understand that there are two ways to match the 50-ohm impedance of the source and the antenna system. You can use an antenna with a 50-ohm impedance and a resonant frequency as close as possible to the working frequency, which means that its impedance will be 50 ohms at that frequency ("radiation resistance") and no reactance. Or you can measure the actual antenna impedance at the desired working frequency and add a matching network.

There is one important point to understand here: Both solutions are effective. Even a trumpet is a perfect antenna at any frequency with the proper matching network (assuming it is a metallic trumpet)! Essentially, if it is matched, then the reflected power will be null. And because there aren't any dissipating resistors in a trumpet, all the applied power will need to go somewhere, which means it will be radiated in the air with no loss. There are only two issues: The first is that the radiation pattern of a trumpet will probably be a little strange. The second is a little more insidious.

Suppose that the impedance of the antenna is very far from 50 ohms and very reactive, say $1 - j \times 1000$ ohms, which is not uncommon for nonoptimal antennas. To match it and cancel the 1000-ohm capacitive reactance, you need to add a serial inductor with a reactance of 1000 ohms. Unfortunately, you can't buy an ideal inductor. Good inductors may have a quality factor Q of, say, 200, which means that their parasitic resistance will be 200 times less than their reactance (here $1000/200 = 5$ ohms). If you compare this value with the 1-ohm resistance of this antenna at resonance, you can see that five times more energy will be dissipated in the matching inductor than radiated by the antenna, which is not optimal. This is only an example, but it is significant. If you have an antenna with an impedance far from the source impedance, building a matching network will often be practically impossible, even if it is always theoretically possible.

Basic Antennas

The easiest solution is then to try to use an antenna that provides an impedance of around 50 ohms and no reactance at the desired frequency without any need for a matching network. Two basic antenna designs have this characteristic and are often used. The first one is the half-wave dipole, which has an impedance of 68 ohms—not far from 50 ohms (see Figure 14.1a).

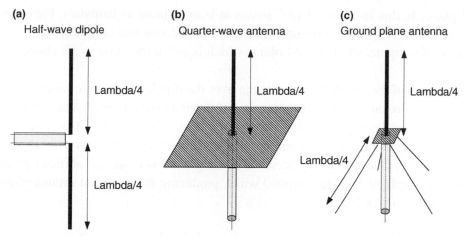

(a) Half-wave dipole **(b)** Quarter-wave antenna **(c)** Ground plane antenna

Figure 14.1: The simplest antenna is the half-wave dipole (a), which is made by two lambda/4 wires driven by a bipolar source. Take half of it, add a reference ground plane, and you have a quarter-wave antenna (b). Finally, if you don't have a good ground, you can add one or several artificial ground wires for the so-called ground plane antenna (c).

The half-wave dipole is built with two collinear wires, each lambda/4 long, with lambda the wavelength in free space of an electromagnetic wave at the desired frequency. As a reminder, lambda is calculated as $\lambda = C/F$, with lambda in meters and F in hertz; C is the speed of light (300,000,000 m/s). The examples in this chapter are based on 868-MHz antenna designs. (868 MHz is a free European band for industrial, scientific, and medical (ISM) applications, somewhat equivalent to the 915-MHz band in the United States.) Let's calculate the wavelength in free space at 868 MHz:

$$\lambda = \frac{C}{F} = \frac{300,000,000}{860,000,000} = 34.56 \, \text{cm}$$

A dipole at 868 MHz will need two 8.65-cm wires (i.e., 34.56/4). Both wires need to be driven by a bipolar electrical signal. A common variant of the dipole is the quarter-wave antenna (see Figure 14.1b), which is derived from the half-wave dipole. The median plane of a dipole is always at a null potential, so it can be replaced by a ground plane. An 868-MHz quarter-wave antenna is thus an 8.65-cm vertical wire above a large ground plane. In this instance, "large" means at least as large as lambda/4, for example, 10×10 cm. Such a quarter-wave antenna has an impedance that is half of the impedance of a dipole, so $68/2 = 34$ ohms, which is still quite close to 50 ohms.

The advantages of the quarter-wave antenna over the dipole are that a quarter-wave antenna can be driven by a unipolar signal (e.g., from a coaxial cable) and that it is physically smaller. Its major disadvantage, which is often forgotten, is that a quarter-wave antenna works only with a good ground plane, which is not an easy thing to get in a portable device. One way to replace a defective ground is to use an artificial ground made with a couple of lambda/4 ground wires, producing the so-called ground plane antenna (see Figure 14.1c).

What about radiation patterns? An antenna that radiates the same power in every direction is called an isotropic antenna. It is used as a reference to compare real antennas. By comparison, the radiation pattern of a half-wave dipole or quarter-wave antenna is like a donut, with a null field in the axis of the antenna and a maximum field perpendicular to it. In this perpendicular direction, the radiated field is 2.15 dB higher for a dipole than for an isotropic antenna. Because the dipole is concentrating the power in this direction, a dipole antenna is said to have a 2.15-dBi gain, which means 2.15 dB more than an isotropic antenna. In real life, things are a little more complex, but that's the idea. Why not simulate it?

Antenna simulation doesn't have to be expensive. Let's start with a free simulator, the NEC2, which began as a U.S. military research project and is now in the public domain, at least in its second version. Arie Voors developed a good user interface for NEC2 with 3D features. It's called 4NEC2. The simulator needs you to model your antenna and all its surroundings using a mesh of round wires. This can be tedious, but it can be automated with scripting routines. Figure 14.2 shows the design of a quarter-wave 868-MHz antenna placed on a ground plane. I simulated the ground plane with four horizontal wires interconnected by a "square" of wires, which itself is 30 cm (around 1 ft) above ground level.

Some more clicks and NEC2 will show you the reflection coefficient of your antenna, which helps ensure that it is well matched at 868 MHz (see Figure 14.3). More impressively, 4NEC2 can also show you the radiation pattern of your antenna in a 3D

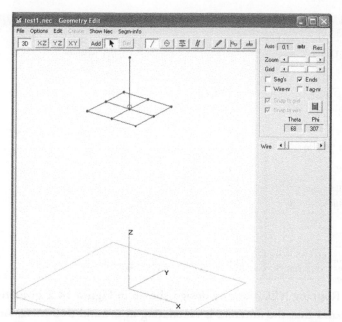

Figure 14.2: This is a quarter-wave 868-MHz antenna on a small ground plane designed with 4NEC2. The generator is the little circle at the bottom of the antenna. The ground plane is 30 cm (approximately 1 ft) above ground level.

view (see Figure 14.4). Designing and executing this simulation took me about two hours without any prior experience with NEC, so don't be afraid to try it.

Are you going to present an antenna design to a customer? Even if your proposal is perfectly matched, I bet you will always get this request: "Can we have a smaller one, please?" What happens if you start with a quarter-wave antenna and reduce its length without changing the operating frequency? Its radiation resistance will continue to decrease, and its reactance will become increasingly capacitive (meaning negative). Reciprocally, an antenna longer than lambda/4 will have a higher radiation resistance and an inductive reactance (like a long wire), so in both cases it is no longer matched.

You can add a matching network to correct it. If the antenna is just slightly shorter than lambda/4, you can use the simplest matching network, which is a serial inductor. The

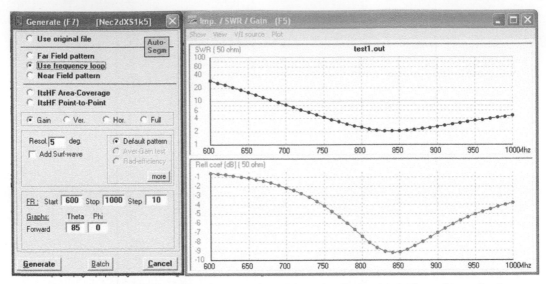

Figure 14.3: Running NEC2 on the design shown in Figure 14.2 gives this reflection coefficient. It shows that the antenna is resonating at around 845 MHz. At 868 MHz, the S11 coefficient is −8 dB. For 1 W of applied power, you have 10 − 8/10 = 0.15 W reflected back to the source. That's 0.85 W transmitted. Not too bad.

inductor will cancel out the capacitive reactance, but it won't increase the radiation resistance, so the match won't be perfect. The matching inductor is often made with a small helical section in the middle of a vertical antenna. If you want to have short antennas or avoid a matching network, the other solution is to use more sophisticated antenna designs.

Microstrip Antennas

Let's assume that you only have 3 cm for your 868-MHz antenna, which is significantly less than the required 8.65 cm. One way to have a smaller antenna is to build it on a material other than air. The length of a resonant antenna is always lambda/4, but the wavelength lambda of a wave is smaller on a substrate (e.g., a ceramic or even a classic FR4 PCB). The ratio is the inverse of the square root of the dielectric constant of the material (which is 4.3 for FR4), at least if the antenna is buried inside the material.

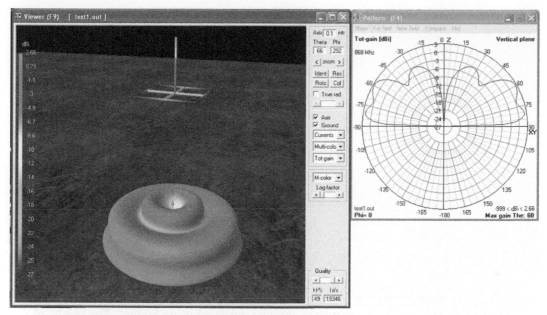

Figure 14.4: The 4NEC2 also enables you to plot the radiation pattern of the antenna, either on a graph or on a 3D plot. Here, the classic donut pattern of a quarter-wave antenna is perturbed by the proximity of the ground. The gain in the privileged direction is around 3 dBi. Greyed on the antenna wires are an image of the RF current circulating in them.

Therefore, if your 868-MHz antenna is a copper track on a PCB (a microstrip antenna; see Chapter 2), it will be significantly shorter than the 8.65 cm calculated earlier. The actual length is difficult to calculate because the antenna is at the interface between epoxy and air, but this is one way to reduce its size. That's why antennas for mobile phones are often built on exotic ceramics.

The bad news is that free simulators, such as NEC, can't help for such antennas. A real electromagnetic simulator is needed. I used Sonnet for the examples below. Sonnet has a key advantage. There is a free version, Sonnet Lite, which you can download from the Sonnet website. Some features aren't included in the free version: Radiation pattern calculation is not supported and design complexity is limited. Still, this is a great

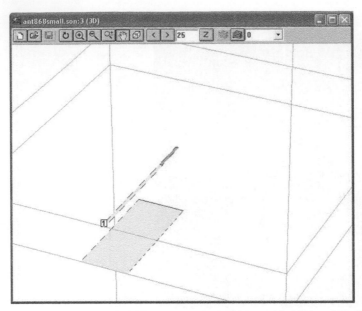

Figure 14.5: This is a 3-cm microstrip antenna designed using Sonnet. The height of the PCB is 1.6 mm, but it is enlarged on the display. Its bottom side holds a ground plane up to the start of the antenna section. On the top side, the track width is calculated for a 50-ohm impedance. This antenna is far too short for 868 MHz.

starting tool. I was fortunate to use the full version (courtesy of Sonnet) for this work. Figure 14.5 shows a first try, with a straight, 3-cm antenna built on FR4 epoxy.

Unfortunately, the electromagnetic simulation shows that the antenna's resonant frequency is still too high, which is around 1520 MHz. Three centimeters is definitely too short for a single antenna, even on FR4. At 868 MHz, the calculated impedance of this small microstrip antenna is 12–98j ohms. One possibility is to add a discrete matching network, which can be calculated as explained in Chapter 1. In that case, the matching network could be a 6.5-pF parallel capacitor and a 21.8-nH serial inductor, as shown in Figure 14.6. Launch the Sonnet simulator one more time and it will show you that the matching is achieved at 868 MHz as planned (Figure 14.7).

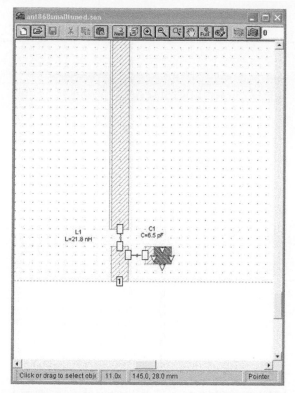

Figure 14.6: A matching network can be designed and simulated using Sonnet.

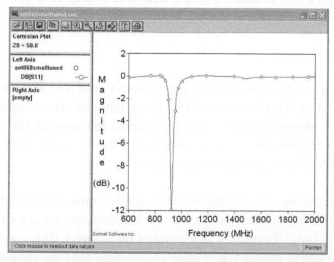

Figure 14.7: The simulation of the short 3-cm microstrip antenna and its matching network helps to ensure that a proper resonance is achieved at 868 MHz. Power transfer will be nearly perfect—at least if the matching network is made with perfect components.

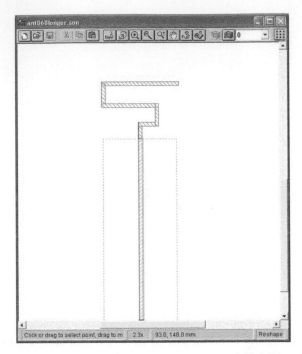

Figure 14.8: This is a first attempt to have a resonance at 868 MHz with an antenna that is still 3 cm long. The length of the _S_ is tuned to get a resonance (see Figure 14.9).

Another solution is to extend the length of the antenna until it is long enough to get a resonance at 868 MHz. That is what I did in the example shown in Figure 14.8. The simulation result is shown in Figure 14.9. The antenna resonates at 868 MHz, even if it has other resonant frequencies. The only issue is that the antenna's calculated radiation pattern is a little chaotic, as its shape might imply.

The beauty of electromagnetic simulators is that it takes only a matter of minutes (at least on a fast PC) to try different solutions—for example, to match a short antenna, you can also add a parallel capacitor at the end of the antenna. This can be done by adding a large surface of copper (see Figure 14.10). This time the simulated response was nearly perfect and its radiation pattern was more acceptable (see Figure 14.11). You have your antenna.

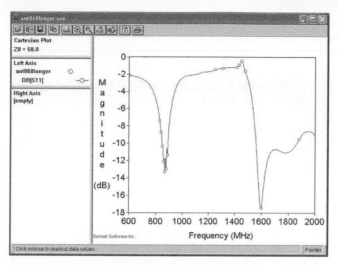

Figure 14.9: I used Sonnet to calculate the reflection coefficient of the antenna in Figure 14.8. There is a resonance at 868 MHz. There are also resonant frequencies of around 1600 MHz, probably due to the shorter elements of the S. However, the radiation pattern of such an antenna (not shown) is disappointing.

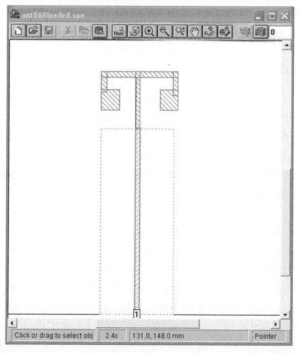

Figure 14.10: Another way to reduce the resonant frequency is to add loading capacitors to the end of the antenna. A capacitor is nothing more than some more copper. I tweaked the size of both squares until I obtained a good resonance frequency (see Figure 14.11).

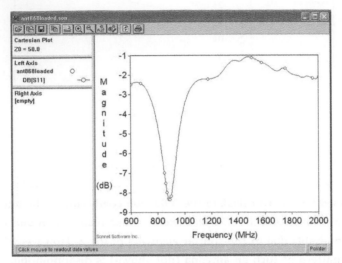

Figure 14.11: This is the simulated reflection coefficient of the antenna in Figure 14.10. It shows a good match at 868 MHz and no spurious resonances at frequencies of up to 2 GHz. Moreover, the antenna's radiation pattern is acceptable.

Antennas in Real Life

Simulations are great tools, but I must be honest: Real life is often more difficult than simulations, in particular with antennas. One of the trickiest points is that each time you move an object close to an antenna, its characteristics can change, so it can be detuned. "Close" here means closer than around lambda/10, where lambda is, as usual, the wavelength in free space. That's why you can design a decent antenna and get poor results when you put the device in a box.

So how are you going to manage the antenna in your next wireless project? If you are in a hurry, take the easy route. First, limit yourself. Design simple and predictable antennas such as quarter-waves or use prebuilt antennas such as ceramic patches and use them exactly as described in their specification. Next, add a large ground plane or use artificial grounds (except when you have a dipole antenna driven by a pair of differential signals, which doesn't need a ground) and connect the ground plane to the electrical ground in a proper RF way (not by a long wire, which acts as an inductor, but with a large contact). Finally, keep the antenna far enough away from any materials. It should work reasonably well.

These rules are not fun, and unfortunately they are often inapplicable because of space or design constraints. Another option is to design your custom, properly matched antenna on a simulation tool, either wired or on a PCB, and optimize it until you have good simulated results. Tools such as NEC and Sonnet will definitely help.

There is another mandatory step, but you may need deeper pockets. You will have to measure whether your antenna is behaving as planned, especially if it is boxed in a package. The instrument needed to do that is a vectorial network analyzer (VNA), which plots the actual amplitude and phase of the reflected signal for each frequency and enables you to easily tune it or measure its impedance to calculate a matching network. I wrote an article about the design of a low-cost vectorial analyzer, which

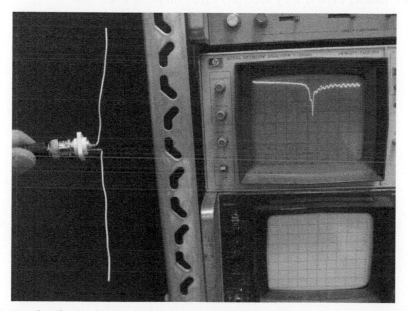

Figure 14.12: I hooked a hand-made 868-MHz dipolar antenna to my old HP8754 VNA. The equipment is invaluable for antenna work because it gives a real-time measurement of the antenna impedance matching. The horizontal scale is 0 to 1300 MHz. The vertical scale is 2.5 dB/division and shows the reflection coefficient of the antenna over frequency. If you buy an old VNA, don't forget to buy the mandatory reflection bridge, which is an HP8502A for the HP8754.

unfortunately lacked the directional bridge needed for antenna measurements ("Vector-SoC," *Circuit Cellar*, no. 149, 2002). Some old VNAs are available on the surplus market at a reasonable price. Refer to Figure 14.12 to see how an old VNA can make your life easier.

Wrapping Up

Here we are. I have touched on only 5% of the subject. I didn't cover several important topics (e.g., broadband antennas; polarization issues; loop, helical, or directional antennas; standing waves; path loss), but I hope that you are now a little less afraid of impedance matching and antennas. Don't hesitate to play with QUCS, NEC2, Sonnet Lite, and real antennas. This is the only way to understand what's going on. Antennas are not black magic, even if they are sometimes on the darker side.

Part 6
Power

CHAPTER 15

Low-Power Techniques: Build Better Energy-Saving Systems

Simple calculations can generate impressive figures, especially when calculating energy waste. According to *Wikipedia*, our worldwide energy consumption is around 15 tera-watt in 2007, which is 15,000,000,000,000 watt. Considering that there are around 6.5 billion people on earth, this is an astonishing number. If we assume that everyone is using the same amount of power, including developing countries, this translates to an average of 2300 W per person. Think about it: The average four-person family (worldwide) uses nearly 10 kW, 24 hours a day, 365 days a year! This is, of course, not only your personal energy consumption but also the energy used to manufacture the goods you buy, the food you eat, the gas in your car, and so on. But that's still impressive, isn't it? And that's for the average human. I would guess that the readers of this book are significantly above average when it comes to energy consumption, right?

The purpose of this introduction is to remind you that we are sometimes unreasonable, and to introduce the subject of this chapter. I will present some tips and ideas that will help you build better energy-saving electronic devices. Let's stay modest, but assuming that better engineering practices can reduce our energy bills by 0.1%, that's still 15 giga-watt worldwide, which is equivalent to a significant number of nuclear plants.

Energy savings are even more critical in battery-operated equipment. The same amount of energy can be wasted in battery- and line-powered products, but in the former case it can be disastrous if you need to replace or recharge the batteries too often. Moreover, in

Note: This chapter is a corrected and improved reprint of the column "The darker side: Low power techniques," *Circuit Cellar*, no. 213, April 2008.

© Elsevier Inc.
DOI: 10.1016/C2009-0-20196-6

Figure 15.1: Here is my imaginary product, a cooking timer with a built-in
oven temperature probe called SmartyCook.

the case of primary (nonrechargeable) batteries, this also implies a major environmental
impact.

SmartyCook

Because the same engineering principles can be used to build green line-powered
products or energy-efficient battery-operated systems, I will use a battery-based
imaginary example for this chapter. Let's assume that you are asked to design the
SmartyCook (see Figure 15.1), an innovative cooking timer with a remote thermal
probe and zillions of new and in-demand software features such as automated *foie gras*
cooking. Sorry for this example, but I have two excuses: I'm French and my wife is a
wonderful *foie gras* chef!

As you know, you can design such a product quickly with a microcontroller. Imagine
that one of your colleagues has already designed the schematic shown in Figure 15.2.
He has found a nano-power PIC variant with an on-chip LCD interface, the Microchip
Technology PIC16F914, which can drive a custom LCD glass. The oven temperature is
measured with a 1000-ohm platinum resistance thermometer, RT1, powered through the
RD1 software-controlled pin and R4 resistor, and fed back to one of the PIC's ADC
inputs. A reference voltage is measured on another ADC input to counterbalance voltage
drops. A small loudspeaker is connected to an I/O pin. Three push buttons are wired to
the RB port, which has interrupt-on-change and internal pull-up features. Finally, three

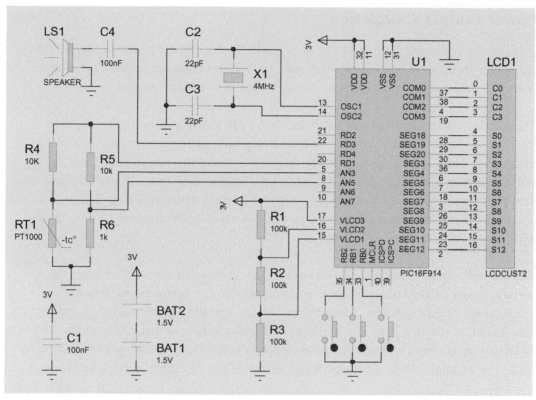

Figure 15.2: Your colleague designed the SmartyCook around a nanoWatt Microchip Technology PIC16F914 microcontroller, a custom LCD glass, a platinum thermal sensor, a loudspeaker for an alarm, and a pair of AAA batteries. Is it a power-optimized design?

resistors, R1 to R3, provide the LCD driving voltages, a 4-MHz crystal is included, and the device is powered by two standard AAA batteries.

Your colleague even told you that he developed power-optimized firmware, which keeps the PIC in low-current standby mode until it is activated by the user. Your task is to check this design on the energy-saving side and improve it, if possible. Caution: Don't ask me for the software and don't complain if the schematic is wrong somewhere. It was just drafted as an example for this chapter. It was never built.

Power Budget Calculation

How would you start? You first need to evaluate the battery life of this design to find out which areas must be improved and how far you are from the target. This is the energy budget calculation, which is a four-step process for any project.

The first step is to identify what I call the *device operating modes*. Each operating mode corresponds to one, and only one, instantaneous current consumption. For the SmartyCook example, I have identified three basic operating modes (see Table 15.1a): standby, active, and beep. You also need to estimate the time spent in each mode every day. This may be more or less difficult, but you do need estimates. Here I have assumed that the cooking timer is used on average twice per day, each time for 15 minutes and with 30-second loudspeaker activation.

The second step is to calculate or measure the *instantaneous current*, which is the current drawn on the battery in each operating mode. The components' datasheets are good starting points. Don't forget to read them completely because the current announced on the first page is often described under very optimistic conditions: no watchdog active, no (or few) internal peripherals active, lowest supply voltage, and so forth. For example, look at the operating current of the PIC16F914 in Figure 15.3.

Also, don't forget to include the currents going through any wire or passive component of the design, including pull-ups or imperfect capacitors. The best way is to check it systematically. Go through the schematic wire per wire for each operating mode and ask yourself if any current can go through it. The result is Table 15.1b, with instantaneous currents going from 12 µA in standby mode to up to 10 mA when the loudspeaker is active.

The third step is the easiest. Calculate the average *daily energy consumption*, in mill-amperes per hour, by multiplying each instantaneous current by the time spent in this operating mode and summing all of the modes. Take care of the units: Seconds must be converted to hours and microamps must be converted to milliamps (see Table 15.1c). In my example a little less than 1 mAh is needed each day, which is roughly split as 50% for the active mode and 20% when the loudspeaker is on. The remaining 30% is devoted to standby mode.

Table 15.1 Four Steps to Evaluate Battery Life of the SmartyCook

(a) Operating modes			
Mode	Description	Duration (s/day)	Comment
Standby	Device not used, LCD off, awaiting key press	84,600	24 h minus active
Active	Device in use, temperature measurement, LCD on, count down	1,800	2×15 min
Beep	Loudspeaker active	60	2×30 s

(b) Instantaneous currents (μA)						
Mode	PIC16F914	LCD	R1/R2/R3	R4/R5/R6/RT1	Speaker	Total
Standby	2.150		10.000			12.150
Active	480.000	20.000	10.000	545.455		1,055.455
Beep	480.000	20.000	10.000	545.455	10,000.000	11,055.455

(c) Daily energy requirements					
Mode	Duration (s/j)	Instantaneous current (μA)	Energy (mAh)	% total	
Standby	84,600	12.150	0.285525	28.6	
Active	1,800	1,055.455	0.527727	52.9	
Beep	60	11,055.455	0.184258	18.5	
Total per day (mAh/j)			0.997510	100.0	

(d) Battery life estimation	
Power source type	AAA zinc/Carbon primary batteries
Supplier and reference	Eveready 1212

Capacity (mAh)	Theoretic	Design	Derating
Theoretical capacity (mAh)	464		
Minimal voltage (V)	1.0	1.2	20%
Lifetime (year)	0.1	2.0	20%
Supplier to supplier variation			20%
High currents (mA)	16.0	10.0	−5%
Minimal temperature (°C)	20.0	20.0	0%
Estimated capacity (mAh)			249
Average daily energy consumption (mAh)			0.998
Estimated battery life (j)			250

(a) Identification of the different operating modes. (b) Calculation of the instantaneous current for each operating mode. (c) Summation of the daily energy requirements. (d) Battery life assessment.

	DC CH		Standard Operating Conditions (unless otherwise stated) Operating temperature −40°C ≤ TA ≤ +85°C for industrial −40°C ≤ TA ≤ +125°C for extended					
Parameter No.	Device Characteristics	Min.	Typ†	Max.	Units	Conditions		
						V_{DD}	Note	
D010	Supply Current (I_{DD})	—	13	19	µA	2.0	Fosc = 32 kHz	
		—	22	30	µA	3.0	LP Oscillator mode	
		—	33	60	µA	5.0		
D011*		—	180	250	µA	2.0	Fosc = 1 MHz	
		—	290	400	µA	3.0	XT Oscillator mode	
		—	490	650	µA	5.0		
D012		—	280	380	µA	2.0	Fosc = 4 MHz	
		—	480	670	µA	3.0	XT Oscillator mode	
		—	0.9	1.4	mA	5.0		
D013*		—	170	295	µA	2.0	Fosc = 1 MHz	
		—	280	480	µA	3.0	EC Oscillator mode	
		—	470	690	µA	5.0		
D014		—	290	450	µA	2.0	Fosc = 4 MHz	
		—	490	720	µA	3.0	EC Oscillator mode	
		—	0.85	1.3	mA	5.0		

*Characterized but not tested.

Figure 15.3: This extract from the PIC16F914 datasheet (courtesy of Microchip Technology) shows you the operating current of the chip depending on the supply voltage, clock frequency, and operating mode. Here the voltage is 3 V and the clock is 4 MHz, giving an impressively low current of 480 µA.

Finally, you can calculate the *lifetime of the batteries* in days by dividing their capacities by the previously calculated daily energy requirement. The only issue is that battery suppliers don't provide actual capacity for your application; they provide only "test-bench" capacities in usually optimistic situations. Therefore, you must include derating factors. In particular, you must include the effect of the lowest voltage you can tolerate, which is often higher than the supplier's minimum specified decharge voltage. You also need to take care of the battery's autodischarge for long-life products, the minimal temperature at which your design may survive, the battery supplier–to–battery supplier capacity variation (because your customer may not buy the best batteries), and the effects of current pulses that can degrade the battery faster than planned. These parameters are often unknown and need to be estimated, but an overall 50% derating is not uncommon.

So, starting with a pair of fresh 464-mA AAA zinc/carbon batteries, the estimated lifetime of the SmartyCook with your colleague's design is 250 days. Not too bad, but let's assume that you need a longer battery life. Remember that the daily energy requirement is the sum, for each operating mode, of the instantaneous current multiplied by the time spent in this mode, including standby mode.

In a nutshell, the daily energy consumption of a device is the sum of the energies used in each operating mode:

$$ENERGY_{DAILY} = \sum_{ALL\ MODES} (TIME_{MODE} \times I_{MODE}) + TIME_{STANDBY} \times I_{STANDBY}$$

This will enable you to identify four different ways to save energy: Reduce the operating current (I_{MODE}), reduce the time spent in each power-hungry mode ($TIME_{MODE}$), reduce the standby current ($I_{STANDBY}$), or introduce new operating modes.

Reducing Operating Current

The first idea is to reduce the current supplied by the energy source when the device is operating, and that's a good idea. How do you do that? The first good option is to reduce the power supply voltage. If you look at the PIC16F914's datasheet in Figure 15.3, you will find that its typical supply currents are 40% to 60% lower at 2 V than at 3 V, and that's often the case! For example, the PIC16F914's operating current in the 4-MHz EC mode goes down from 480 to 280 µA. Moreover, all resistive losses are also reduced at lower voltages. You can get this improvement either with a lower voltage battery (but 2 V is not easy to get) or with a low-loss linear regulator, such as the MCP1700 from Microchip Technology, which has a self-consumption of less than 2 µA. Adding this regulator to the original schematic will enable you to extend the battery life by 30%, with a daily energy requirement down from 1 to 0.73 mAh. This is a good start with such a small change!

Don't be confused about such linear regulators: Energy will be lost in the regulator, but globally the current drawn on the batteries will be reduced so that the overall energy used will still be lower than without a regulator. For more complex designs, you can check if a small, low-power, step-down DC/DC converter can help. For example, a Maxim Integrated Products MAX1556 doesn't need more than 16 µA to operate. You can also use a lower-voltage battery and a step-up DC/DC converter working with very low start-up voltages, such as the MAX1674, which starts at 1.1 V, because this solution can also enable you to use the batteries more deeply.

Clock frequency is another good source of savings because any CMOS circuit has power consumption that is roughly proportional to its operating frequency. Here your colleague has planned a 4-MHz crystal probably because it is a common value for microcontrollers, but do you think such a design actually needs so many MIPS? Because the LCD is managed in hardware, you need to count down seconds and measure a thermal probe from time to time so you can replace the 4-MHz oscillator with a 32.768-kHz watch crystal and clock the PIC at that frequency.

The good news is that the PIC operating current goes from 280 to 13 µA, providing another 10% improvement on the overall bill—down to 0.59 mAh/day. Even if your design can't tolerate such a low clock frequency, you can still try to reduce the clock frequency when you don't need the full power, thanks to the on-board clock prescalers or phase-locked loops that are available in many chips.

Finally, check each component systematically and think twice. Is there a lower-current alternative? Look again at Figure 15.2. There are three easy improvements. First, your colleague used a PT1000 platinum sensor, which has a 1-kohm resistance at ambient temperature. If you replace it with a 10-kohm sensor, namely a PT10000, the measurement current will be divided by 10. The same goes for the R1/R2/R3 voltage divider used to generate the LCD voltages. Why 100 kohms? Because the PIC input impedance is high enough that you can safely use 470-kohm sensors, with a 5× current reduction. Lastly, the loudspeaker can be replaced with a resonant piezoelectric transducer with the same audio volume, but with a current that is half as powerful.

After these first easy modifications, refer to the schematic in Figure 15.4. It is not very different from the initial one, but your daily energy is reduced from 1 to 0.21 mAh. Another way to express this 80% improvement is that the SmartyCook's lifetime on a single set of batteries is extended from 250 to up to 1153 days as shown in Table 15.2!

Reducing Operating Duration

Energy is proportional to current multiplied by time. You have drastically reduced the operating current, but can you do something about operating duration? One way would be faster processing, but that would require a quicker clock, which means higher currents. An alternative to processing-hungry systems is to select processors with built-in hardware accelerators. For example, a hardware multiplier can give a real

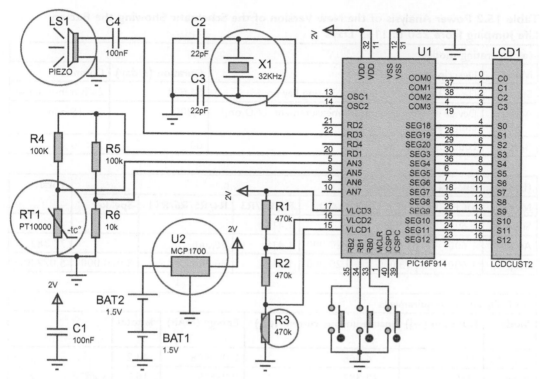

Figure 15.4: This schematic is nearly identical to Figure 15.2. However, its theoretical battery life is four times longer, thanks to reduced operating currents!

energy bonus on DSP-like applications. Don't forget that you can increase battery life just by using a better compiler. A better compiler will give you an optimized code and will not waste energy through extra clock cycles. The difference can be significant. Of course, you can also hand-optimize the critical sections of the code.

But there are plenty of other ideas. For example (even if it is not applicable in this example), you can reduce power requirements at the user interface level. If the user can perform a task quicker with a better interface, then the power consumption will be lower. Do you know how many CPU cycles are needed to dig into a hierarchical menu structure on your phone? Wouldn't it be easier to have one button dedicated to the feature you use the most? In the same spirit, you can also transfer data to external systems more quickly (e.g., when using power-hungry RF links). Data compression can help there.

Table 15.2 Power Analysis of the New Version of the Schematic Showing the Battery Life Jumping from 250 to 1153 Days

(a) Operating modes

Mode	Description	Duration (s/day)	Comment
Standby	Device not used, LCD off, awaiting key press	84,600	24 h minus active
Active	Device in use, temperature measurement, LCD on, count down	1,800	2 × 15 min
Beep	Loudspeaker active	60	2 × 30 s

(b) Instantaneous currents (μA)

Mode	MCP1700	PIC16F914	LCD	R1/R2/R3	R4/R5/R6/RT1	Speaker	Total
Standby	1.600	1.050		1.418			4.068
Active	1.600	13.000	20.000	1.418	36.346		72.382
Beep	1.600	13.000	20.000	1.418	36.346	5,000.000	5,072.382

(c) Daily energy requirements

Mode	Duration (s/j)	Instantaneous current (μA)	Energy (mAh)	% total	
Standby	84600	4.068	0.095608	44.2	STANDBY / ACTIVE / BEEP
Active	1800	72.382	0.036191	16.7	
Beep	60	5,072.382	0.084540	39.1	
Total per day (mAh/j)			0.216339	100.0	

(d) Battery life estimation

Power source type	AAA zinc/Carbon primary batteries
Supplier and reference	Eveready 1212

Capacity (mAh)	Theoretic	Design	Derating
Theoretical capacity (mAh)	464		
Minimal voltage (V)	1.0	1.2	20%
Lifetime (year)	0.1	2.0	20%
Supplier to supplier variation			20%
High currents (mA)	16.0	10.0	−5%
Minimal temperature (°C)	20.0	20.0	0%
Estimated capacity (mAh)			249
Average daily energy consumption (mAh)			0.216
Estimated battery life (j)			1,153

Another trick is to use "wasted" CPU cycles to do something useful. In particular, when using crystal oscillators, there is a frequency stabilization delay that can be counted in milliseconds. Often the boot code of the processor just waits for stabilization. The cycles can be used to do anything you want as long as they are not time-critical, such as variable initializations. If you can accept a little degradation in precision, you can also replace your crystal with ceramic resonators, which usually have a much shorter startup time (see Chapter 9).

Let's go back to the SmartyCook. There is one operating mode where things can easily be done more quickly: alarm beep generation. As part of the hypothesis, I stated that the alarm was 30 seconds long, but nothing prevents you from drastically reducing the alarm's duty cycle. The resonant piezoelectric transducer can be activated, for example, for 100 milliseconds every 2 seconds or so, with the same functionality. It is a small modification, but because the alarm mode was starting to be a significant waste after the previous optimizations, this gives you another 35% improvement, with the estimated battery life jumping from 1153 to 1765 days.

Introducing New Operating Modes

You can also try to break power-hungry operating modes into more optimized modes. The easiest example is when a processor waits for something: the end of a timer, the end of an ADC cycle, or an external event. Power-conscious engineers will never leave the processor in the same operating state, but will use the processor's hardware resources (i.e., interrupts) to put the processor in a low-power idle or standby mode. However, in the SmartyCook design, the gains will be minimal because the system clock is already very low.

Another way to introduce new operating modes is to double-check each component and the wiring of the design. Can it be switched off more often? This approach needs to be fully systematic because you may forget basic components, such as pull-up resistors. Doing this exercise on the already optimized SmartyCook schematic enables you to find two possible improvements on the schematic provided in Figure 15.4. Have you found them? The first one is the R1/R2/R3 network. It is always powered, but it is needed only when the device is operating, not in standby mode, when the LCD is off. Just power it from a free GPIO pin and you can save 1.4 μA. That's a few micro amps, but it is still a 30% improvement on standby current.

The second improvement is related to the thermal probe. Until now I'd assumed that it is powered when the device is operating. But it needs to be powered only when it is measuring the temperature, and this can be as quick as 1 ms each second through the introduction of a new "measurement" operating mode. The schematic in Figure 15.5 sums up all of these small changes. It is still close to the initial one, but the estimated battery lifetime jumps to 2790 days, which is already more than 10 times longer than the original design!

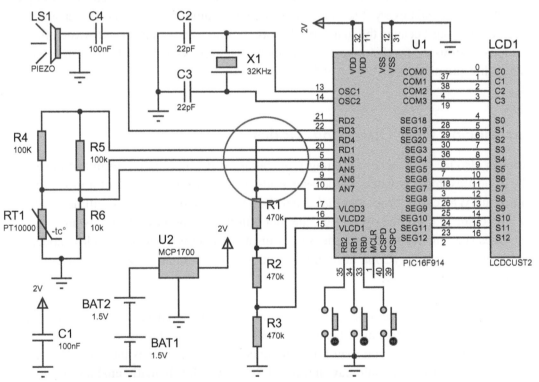

Figure 15.5: After a minor hardware change, just by powering R1/R2/R3 through a GPIO and making a small firmware change in thermal probe management, the battery life is doubled compared to Figure 15.3.

Reducing Standby Current

Now the operating modes are optimized and the standby power consumption is starting to be significant. This is the last savings you need to work on. You have already checked that you are using low-standby current components and you have turned off any unused component or internal peripheral (unused ADCs, timers, clocks, and so on). You have also double-checked that each pin is configured in its lowest current consumption mode before entering sleep mode. It is usually the "output" state, but it can change from one microcontroller to another.

What else is there to do? Because RAM is usually preserved in sleep mode, a processor with a smaller RAM may be more power friendly than a larger one, so don't keep large components if you don't need them. Another trick is that a processor or its oscillator is often kept active in standby mode just to enable a software-based real-time clock to run. This may be a cost-effective solution, but it is usually not power-efficient. Don't forget that a dedicated RTC chip, such as the Maxim Integrated Products DS1302, needs just 300 nA at 2 V to keep the time!

The best way to reduce standby power is to have zero standby power, which usually means a small mechanical on/off switch. That's a very good solution, but you will dislike it because it implies one more action before using the device and because you will forget to switch it off 50% of the time. A better solution is to implement a real "zero-power" mode using one of the device's buttons as a hard-wired power-on. Look at the principle in Figure 15.6.

The SmartyCook design can easily use this trick just by replacing the MCP1700 regulator with a model equipped with a standby input pin (i.e., the MAX1725) and a couple of diodes. The final schematic is shown in Figure 15.7.

Wrapping Up

The SmartyCook is now a little more complex than in the initial design, but compared with the energy analysis in Table 15.3 on page 266, it is impressive. Its theoretical lifetime on two standard AAA batteries jumped from around 250 days to more than 15 years! Of course, these calculations are approximate and would need to be checked on an actual

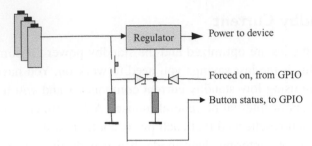

Figure 15.6: This trick enables you to implement a real zero-power standby mode with little end-user impact. In standby mode, the main regulator is switched off until you press a specific button, which pulls the regulator's standby pin high. The main microcontroller is then powered on. The firmware execution starts and immediately switches a GPIO pin to a high-level logic state to maintain the regulator power even if the user releases the button. The device then operates normally and all of the buttons are available to the application until the microcontroller decides to switch itself off by releasing the GPIO pin. Make sure that the processor can tolerate a voltage on the button input pin that is higher than the regulated voltage, which isn't an issue with 5-V-compliant inputs.

prototype. You may also argue that the first design was not well optimized to allow me to write this chapter, and you would be right. Nevertheless, I guess that you wouldn't have bet on an improvement of a factor of 20 when you started reading this chapter.

In conclusion, I want to emphasize that all of the improvements I discussed must work together in a holistic manner. When operating modes are optimized, standby modes start to hurt and vice versa. If you are tracking the micro amp, you must take your time and check everything. I hope this chapter will help you design more energy-saving devices. You may also use it as a checklist for your next project.

Even if the impact on our 15-TW global energy consumption may be limited, it will somehow help our planet. I also hope that you are convinced that low-power techniques are not black magic, even if they are sometimes on the darker side.

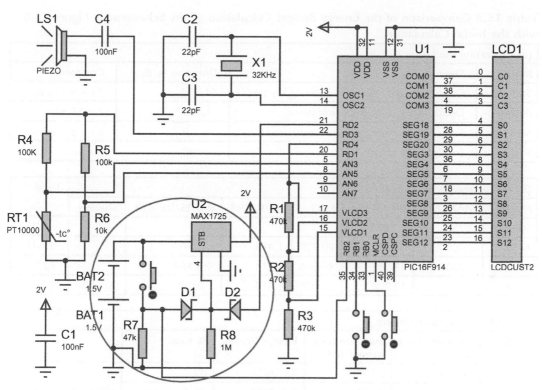

Figure 15.7: This is the final version of the SmartyCook schematic. One more pin on the power supply regulator enables you to implement a real zero-power standby mode, which was presented in Figure 15.6.

Author's Note: I first presented the method explained in this chapter in the French publication, *Ivry-sur-Seine*, on January 10, 2008, during an ecodesign seminar that was organized as part of the ECO'TRONICS initiative. This program is managed by Jessica France and financed by French state and regional public funds. I want to thank them for their support and action for better environmentally friendly devices.

Table 15.3 Comparison of the Energy Budget Calculation of the Schematic in Figure 15.6 with the Initial Calculation

(a) Operating modes

Mode	Description	Duration (s/day)	Comment
Standby	Device not used, LCD off, awaiting key press	84,600	24 h minus active
Active	Device in use, LCD on, count down	1,800	2 × 15 min
Measure	Same but with temperature measurement on	9	5 ms per second
Beep	Loudspeaker active	60	2 × 30 s

(b) Instantaneous currents (μA)

Mode	MAX1725	PIC16F914	LCD	R1/R2/R3	R4/R5/R6/RT1	Speaker	Total
Standby	0.700	0.000					0.700
Active	2.000	13.000	20.000	1.418	0.000		36.418
Measure	2.000	13.000	20.000	1.418	36.364		72.782
Beep	2.000	13.000	20.000	1.418	0.000	500.000	536.418

(c) Daily energy requirements

Mode	Duration (s/j)	Instantaneous current (μA)	Energy (mAh)	% total	
Standby	84,600	0.700	0.016450	37.6	
Active	1,800	36.418	0.018209	41.6	
Measure	9	72.782	0.000182	0.4	
Beep	60	536.418	0.008940	20.4	
Total per day (mAh/j)			0.043781	100.0	

(d) Battery life estimation

Power source type	AAA zinc/Carbon primary batteries
Supplier and reference	Eveready 1212

Capacity (mAh)	Theoretic	Design	Derating
Theoretical capacity (mAh)	464		
Minimal voltage (V)	1.0	1.2	20%
Lifetime (year)	0.1	2.0	20%
Supplier to supplier variation			20%
High currents (mA)	16.0	10.0	−5%
Minimal temperature (°C)	20.0	20.0	0%
Estimated capacity (mAh)			249
Average daily energy consumption (mAh)			0.044
Estimated battery life (j)			5698

CHAPTER 16

From Power Line Measurements to PFC

In this chapter I will start by discussing how the power consumption of line-powered devices is measured. This may seem basic and less exotic than usual, but RMS power, apparent power, and power factors are sometimes misunderstood. Moreover, measurement tools are not so easy to use efficiently and, more important, safely. This will lead me to introduce power factor controllers, and I will share with you some experiments on a couple of pretty cool chips. Ready? Let's go.

Some Basics

Assume that you need to measure the power consumption of your latest device, which is powered from a 110-V or 220/240-V line voltage, depending on whether or not you like ketchup. What do you actually need to measure power consumption? Let me remind you of some basic notions about AC power.

If a device is connected to a source providing a voltage $V(t)$, usually a sine wave, and draws a current $I(t)$, then Ohm's law says that the instantaneous power consumption over time is simply the product of $V(t) \times I(t)$:

$$P(t) = V(t) \times I(t)$$

The real power used by the device, also called the active power or RMS power, is then simply the average of this instantaneous power over time, which is, mathematically speaking, the integral that follows.

Note: This chapter is a reprint of the column "The Darker Side: From power line measurements to PFC," *Circuit Cellar*, no. 229, August 2009.

© Elsevier Inc.
DOI: 10.1016/C2009-0-20196-6

$$P_{real} = \frac{1}{T} \int V(t) I(t) \, dt$$

When the instantaneous voltage and power are measured at discrete time steps and not continuously, which is usually the case in actual electronic systems, this integral can be simply rewritten as a discrete average of the voltage samples multiplied by the current samples, i.e., the sum of the $V \times I$ samples divided by the number of samples:

$$P_{real} \approx \frac{1}{N} \sum_{i=1..N} V_i \times I_i$$

First difficulty: This average must be calculated over exactly one or more full AC periods to avoid measurement errors. Of course, the number of sampling points per period must be large enough for proper averaging. Thanks to the Shannon theorem, you know that this means more than twice the frequency of the highest significant harmonic (as we will see, the current is usually not a sine wave), which usually translates into a couple of thousand samples per second.

The AC voltage and AC current can also be independently measured through their RMS values, which are calculated as follows:

$$V_{RMS} = \sqrt{\frac{1}{N} \sum_{i=1..N} V_i^2}$$

$$I_{RMS} = \sqrt{\frac{1}{N} \sum_{i=1..N} I_i^2}$$

Once again, these sums must be calculated on exactly one or more full periods to avoid errors. Usually, where the line voltage is a sine wave of amplitude $\pm V_{PEAK}$, you probably know that the RMS values are simply the peak values divided by the square root of 2. For example, a "220-V" line voltage corresponds to a 220-V RMS voltage or to a sine voltage ranging from −311 to +311 V because 220 × 1.414 = 311.

The ratio between the peak value and the RMS value is called the crest factor. This crest factor is 1.414 for a sine wave. For a stable DC signal, the RMS value is equal to the "peak" value, so the crest factor is 1, which is the minimum. However, it can be much higher for pulse-type signals.

For example, if the current is 10 A during 2 μs and then 0 during the remaining 19.998 ms per 50-Hz period, the peak current will be 10 A, but the RMS current will be the square root of $10^2/10,000$, which is 0.1 A, giving a crest factor of $10/0.1 = 100$. Therefore, the crest factor is a good measurement for evaluating the shape of the instantaneous current, voltage, or power. Figure 16.1 summarizes the different scenarios.

What can you do with the RMS voltage and current? Of course, you can multiply them, and this will give you the apparent power used by the device, a quantity usually measured in VA (volt × ampere), in order to differentiate it from the real power, which is measured in watts:

$$P_{apparent} = V_{RMS} \times I_{RMS}$$

This apparent power is always equal to or higher than the real power—trust me or find it out yourself. The ratio of the two is called the power factor, which is always less than 1:

$$PF = \frac{P_{real}}{P_{apparent}}$$

If the voltage is a sine wave and if the load is a pure resistor, the current will be an in-phase sine wave. In that case the apparent power will be equal to the RMS power, giving a power factor of 1. If the load is slightly capacitive or inductive, the current will still be a sine wave but with a phase shift, either a positive or negative one (see Figure 16.1).

The power factor can then be quite easily calculated. It is equal to the cosine of the phase shift between voltage and current, which is why this power factor is also called $\cos(\phi)$.

In fact, this is also nearly the same even if the current wave is more complex than a sine. Imagine that the current is a complex periodic shape. It can be decomposed as a sum of sine waves thanks to a Fourier transform. It can be shown that some energy can be transmitted to the load only when a given harmonic of the current has the same frequency as the given harmonic of the voltage. If the voltage is a pure sine, then the real power is only dependent on the fundamental bin of the current Fourier transform:

Load	Waves	Current crest factor	Powers	Power factor
DC	DC 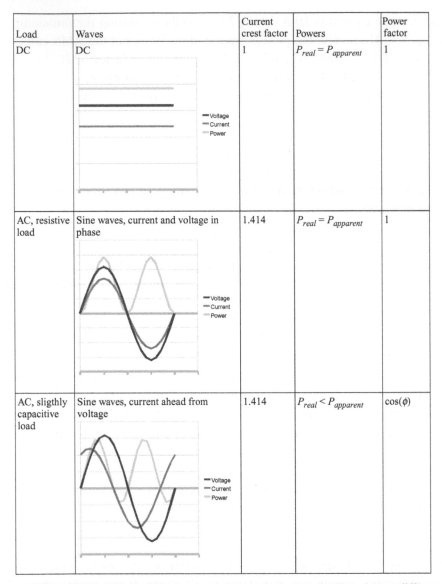	1	$P_{real} = P_{apparent}$	1
AC, resistive load	Sine waves, current and voltage in phase	1.414	$P_{real} = P_{apparent}$	1
AC, sligthly capacitive load	Sine waves, current ahead from voltage	1.414	$P_{real} < P_{apparent}$	$\cos(\phi)$

Figure 16.1: With a DC voltage or a purely resistive load, there is no difference between real and apparent power. The situation is drastically different with more complex impedances.

AC, slightly inductive load	Sine waves, voltage ahead from current	1.414	$P_{real} < P_{apparent}$	$\cos(\phi)$
	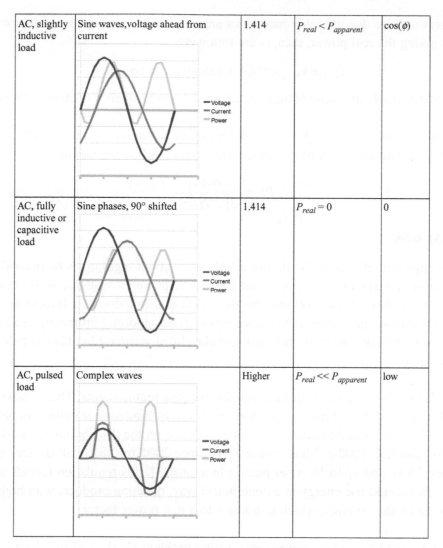			
AC, fully inductive or capacitive load	Sine phases, 90° shifted	1.414	$P_{real} = 0$	0
AC, pulsed load	Complex waves	Higher	$P_{real} \ll P_{apparent}$	low

Figure 16.1: *continued*

All powers that are due to higher harmonics are nullified if you do the calculation. The formula giving the real power, then, is the following:

$$P_{real} = V_{RMS} \times I(fundamental)_{RMS} \times \cos(\phi)$$

The angle ϕ is simply the phase of the fundamental current frequency relative to the voltage.

It can also be shown that the power factor for complex currents is directly related to the total harmonic distortion of the current through the following formula:

$$PF = \frac{\cos(\phi)}{\sqrt{1 + THD^2}}$$

Regulations

Okay, enough with theory, let's talk just a little more about regulations before switching the power on. Energy efficiency is a must for battery-powered systems, as discussed in Chapter 15, but why should you measure the power consumption of a line-powered device? Of course, the answer is to reduce power consumption. Fortunately, we have become more energy-conscious, not only to fight global warming but also to reduce our electricity bills.

But regulators also help us, with increasingly stringent requirements. This is especially true in Europe and Asia. First, we got harmonic current limitation standards to help mitigate EMC issues and to allow efficient energy use on the distribution network. (EN 61000-3-2 and IEC 1000-3-2 has been in force since 2001 for nearly all devices using more than 75 W and up to 16 A per phase.) In a nutshell, this regulation forbids selling products that would use energy in a nonefficient way, meaning products with high levels of harmonic currents, which translate into a low power factor.

It should be noted that a low power factor isn't a problem for the end user because only real power is usually charged by the electricity supplier. But it is a big problem for the supplier because resistive losses in the lines are linked to current ($P = R \times I^2$), whether or not this current is in phase with the voltage. Thus, a low-power factor means the same bill to the user but higher costs for the supplier—now you know why this is regulated. Moreover, harmonic currents are definitely an EMC concern because these high-frequency currents will radiate from the lines.

Standards, such as IEC 1000-3-2, are related to power factor and harmonic distortion, but they don't limit the amount of energy used. Newer standards are emerging that actually limit the energy consumption of devices. In particular, the so-called EuP directive ("Energy Using Products," 2005/32/CE) was signed in July 2005 and took effect in Europe at the end of 2008. One of its targets is to reduce, by no less than 70% in the next 10 years, the overall power used by equipment in standby mode. From January 2010, equipment using more than 1 W in standby mode (2 W if some information is displayed when the device is sleeping, e.g., a digital clock) will be banned throughout the European Community. These limits will be reduced by 50% in 2013 to 0.5 and 1 W, respectively. I bet that the majority of computer power supplies, DVD recorders, or TVs will need to be redesigned, but this will allow us to shut down a couple of nuclear power plants.

Even without mandatory regulations or power-efficiency labels, such as EnergyStar, consumers are becoming increasingly power sensible. Promoting devices that have significantly lower power consumption than those of a competitor is definitely a market advantage not only for green customers but also for businesses.

Measurements

Caution: Even with the recommendations given in this chapter, line voltage measurements are always very risky and can easily be lethal. Never try to reproduce these experiments if you haven't taken a formal course on electric safety rules and if you aren't really experienced.

Okay, now you know why you may need to measure the power of your last line-powered device, but how?

The cheapest test setup is to use a multimeter and to measure the line voltage and line current independently and multiply them (Figure 16.2). Unfortunately, this will allow measurement of only the apparent power, in VA, not the real power, because you will know neither the phase relationship between voltage and current nor the harmonic content of the current (even if the multimeter is able to reliably measure the RMS value of the voltage and current). This may be enough if you are working on a resistive heater, but absolutely not if you have an AC to DC power supply, either switching or linear, as both could have very degraded power factors.

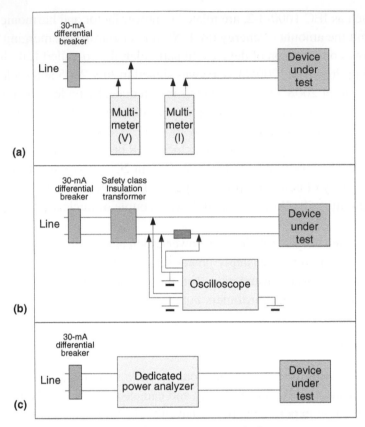

Figure 16.2: (a) Using a simple multimeter to measure the power consumption of a line-powered system allows measurement of the apparent power but not the real power. (b) An oscilloscope can be used but only with great care. In particular, a good insulation transformer is mandatory even if you have an insulated differential probe. (c) A dedicated power analyzer makes life easier.

Without dedicated test equipment, the only other solution is to use an oscilloscope, especially if you are lucky enough to have a digital model with features such as curve multiplication and measurements. It may seem easy: Get voltage on channel 1 and current on channel 2 through a voltage measurement on a shunt resistor or a current clamp; ask the scope to calculate ch1 × ch2, which is the instantaneous power, and calculate the average value over an integer number of periods. Unfortunately, your scope will explode if you don't take some basic precautions—and if you aren't already dead.

Take care because this measurement is REALLY dangerous, and you must think twice before connecting your probes to any line-powered system. The issue is that the inputs of a classic scope are referenced to the scope chassis, which is itself grounded. If you connect the ground tip of the probe to a line voltage, you have a perfect short circuit between the line and the ground through the scope. This measurement setup is possible but must be managed with great care and with at least four mandatory precautions (see Figure 16.2b).

First, you must absolutely use an insulation transformer between the line and the device being tested, and this insulation transformer must be double-insulated. The only alternative is to use an isolated differential probe; however, that's more risky. Still, I recommend the insulation transformer. Second, you must double-check that your scope is grounded and that your installation is protected by a good differential safety breaker (30 mA or even less). Never disconnect the ground of the scope to make it "floating," or all of its metallic parts could become lethal. Third, you must double-check the maximum voltage rating of your probes and take care to ground them to a single point to avoid dangerous ground currents. Finally, the basic electric safety rules apply: Never do such tests when you are alone in your lab; always keep one of your hands in your pocket as long as the power is on; and switch power on only when you are quite far from the device.

The other solution is more expensive but much simpler: Use dedicated test equipment, namely an energy analyzer (Figure 16.2c). You may find very-low-cost domestic wattmeters for less than $20, but these are definitely not measurement grade and will not give you the details you need, such as harmonic contents or U/I waveforms, to optimize your products. You will find professional energy analysis systems from companies such as Fluke (model 43B is a classic), Yokogawa, or Voltech. These systems are basically insulated U/I dedicated scopes with on-board measurement features. I am using an intermediately priced solution, the PowerSpy plug (Figure 16.3), which was developed by Alciom.

Well, Alciom is my own company, so this is the free advertisement section. Technically speaking, however, PowerSpy includes a nonisolated fast measurement circuit and

Figure 16.3: The PowerSpy is a power analyzer based on a small plug-type measurement system linked through Bluetooth to a PC hosting real-time waveform calculations and user interface.

minimal on-board real-time processing mainly for triggering. However, the V and I waveforms, which are digitized at 16 Ksps, are transferred in real time to a nearby PC through a Bluetooth link. This wireless connection provides an easy and bullet-proof insulation barrier, and the PC manages the calculations and user interface through the associated software.

To the Bench

Okay, enough commercials. It's time to show you some actual measurements. I could have used any switching power supply–based design as an example, but I used an AC/

Figure 16.4: This is the ST Microelectronics AC/DC power supply that I used for
the tests. From left to right are the input filter, diode bridge, ballast capacitor,
FET switch, high-frequency transformer, and output sections. The feedback
optocoupler is at the bottom.

DC power supply evaluation kit from ST Microelectronics, the EVL6566B-40WSTB
(Figure 16.4), simply because the architecture is documented and easier to understand.
This module is a classic 40-W low-cost flyback converter based on the company's
L6566B integrated circuit, and it is available from Digikey or Mouser. Figure 16.5
shows you the generic architecture of such a switching power supply, but you can look
at the schematics of this board on the ST Microelectronics website.

I connected the line input of this power supply to the PowerSpy, then connected an
electronic dummy load set at 1.7 A to one of the outputs, and switched everything on.
Voila, I got Figure 16.6. As you can see, the current wave is far from sine-shaped, and
this translates into a very low power factor. With 1.7 A on 9 V drawn from the output
(15.3 W), the real power used from the line was 19.2 W, which provides a reasonable
efficiency of 80%. However, the apparent power was 44 VA, giving a power factor of

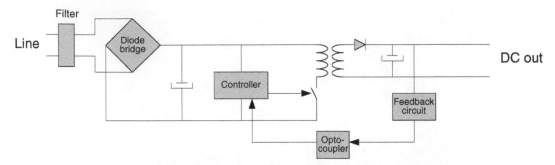

Figure 16.5: Classical architecture of a flyback isolated AC/DC converter. Line voltage is first transformed into a high-voltage DC then switched to a high-frequency transformer through an integrated controller. Output voltage is regulated through a feedback loop usually made of zener diode and optocoupler.

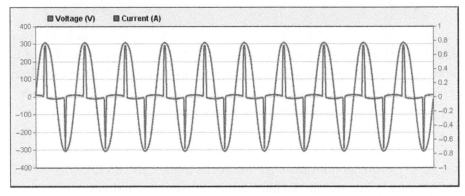

Figure 16.6: The measurements done on the AC/DC power supply with a 15-W load at the output show a very peaky current, which is due to the diode bridge and large ballast capacitor. You can also see a small sine current in quadrature with the voltage because of the input filter of the design.

less than 0.45! Now you understand better why you should not use a multimeter to evaluate the power consumption of such a device: You will be 50% off. The current harmonics are also impressive (Figure 16.7).

Don't be confused—the L6566B-based power supply is not bad. Nearly all simple AC/DC converters will show the same behavior. Why is the current so peaky? Look again

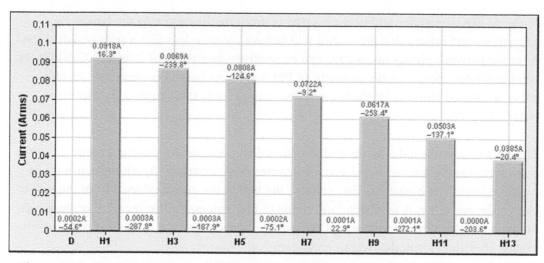

Figure 16.7: A power analyzer such as PowerSpy allows you to display the harmonics of the current through a fast Fourier transform. Here the harmonics are, of course, very high—that is, nearly as high as the fundamental.

at Figure 16.5. The answers are in the second component: the diode bridge and its large ballast capacitor. How does such a bridge work? Simply by allowing the current to flow when the line voltage is above the voltage level of the capacitor. This means that the current from the line will be null except during a small period around the maximum voltage, and this is exactly what is demonstrated in Figure 16.6. The higher the load on the output, the wider the current peak, but this does not in any way significantly reduce the bad power factor on such a simple design.

By the way, the situation is nearly the same with a plain old linear power supply. In that case, the diode bridge will be after a 50/60-Hz transformer, but there will also be power peaks when the line voltage will be higher then the voltage of the ballast capacitor, even if the transformer helps to smooth the current spikes a little.

Such a low power factor is actually not a major issue for low-power systems, and that's why regulations do not yet impose limits on harmonics for products under a given overall power limit, generally 75 W. So what is the solution to higher power supplies? Power factor correctors (PFCs), of course. PFCs are circuits that replace the first stages of the AC/DC converter architecture—that is, from the line input to the ballast capacitor—and ensure that the circuit will behave more or less like a pure resistive load from the network viewpoint. A PFC can be passive—just think of it as an LC 50/60-Hz bandpass filter. However, the majority of PFCs are now based on electronic systems, either dedicated chips or DSPs, because they are cheaper than large coils and more efficient.

The idea is to replace the diode bridge with a direct AC-to-DC switched mode converter, usually a nonisolated boost converter, with a control algorithm that mimics a resistive load: The higher the line voltage, the higher the current drawn from the line to charge the capacitor. A feedback loop needs to be implemented to increase or decrease the value of this virtual resistance if the voltage of the capacitor is too low or too high. Figure 16.8 shows you the overall architecture.

I am sure you want a demonstration. ST Microelectronics has developed dedicated PFC chips such as the L6562A, which is an 80-W pre-regulator PFC single-chip solution. I bought the corresponding evaluation board and hooked it to the AC/DC board used previously, simply by connecting the 400-V DC output of the PFC board to the output of the diode bridge on the AC/DC power supply board. That way, the line voltage is

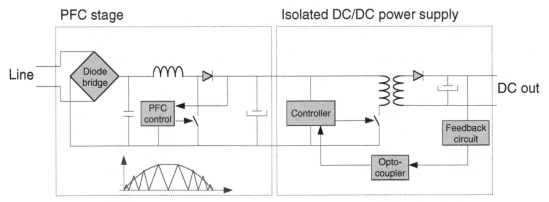

Figure 16.8: A PFC replaces the diode bridge and large ballast capacitor with a smaller capacitor, followed by an intelligent boost converter, which then charges the ballast capacitor. The PFC modulates the switching element to mimic a resistor load, ensuring that the average current is proportional to the line voltage.

Figure 16.9: Here is how I connected the PFC evaluation board to the AC/DC power supply evaluation board. None of the input section of the latter is used any longer. It has been replaced by the PFC controller connected directly to the output of the diode bridge.

first transformed to a high DC voltage, thanks to the PFC board, and then converted back to a low DC voltage through the AC/DC supply board used as a DC/DC converter. Figure 16.9 shows you the setup.

Switch the power on, still with a 15-W load on the output, and you will measure a much cleaner pseudo-sine wave current, as demonstrated in Figure 16.10. The numeric results are also very interesting: With the addition of the PFC stage, the real power increases from 19.2 to 22.8 W. This 18% increase is due to the efficiency of the PFC stage, which isn't 100%, especially with a low 15-W load on the output. The loss would be around 5% with a more usual 40- to 80-W load. However, this small increase in real power is counterbalanced by a drastic reduction in the apparent power from 44 VA without PFC to 23.5 VA with PFC. Similarly, the power factor increases from 0.45 to 0.97! Of course, the harmonic distortion is also impressively improved.

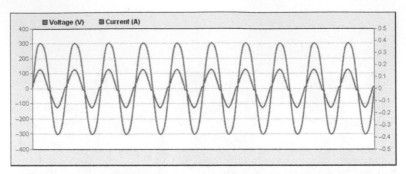

Figure 16.10: With the PFC circuit, the current waveform is drastically improved and very close to a pure sine. Some distortion is visible at the zero crossing but is nothing compared to Figure 16.6.

Shutting Power Down

You will not need a PFC in all your projects. However, you will definitely need to look at these solutions when you build a power supply above 75 W because there are no other solutions for complying with IEC regulations. Moreover, you will definitely increasingly need to optimize all your line-powered devices to satisfy your customer, your regulator, or the planet. I am sure that linear AC/DC power supplies will no longer exist is a couple of years.

Wrapping Up

I hope that you now have a better understanding of line power issues, measurement methods, and constraints. Don't hesitate to play with these converters. High-voltage electronics are fun too. But, once again, please take care—220/240 V is definitely dangerous, but 110 V is also lethal. And capacitors stay loaded for a long time.

Part 7
System Control

CHAPTER 17

PID Control Without Math

The vast majority of real-world systems are based on feedback loops. Broadly speaking, a feedback loop is used to manipulate the inputs to a system to obtain a desired effect on its outputs in a controlled way. Your DVD player uses a feedback loop to drive its spinning motor at a precise rotation speed. Your mobile phone has a feedback loop to adjust its transmission power to the required level. And, of course, your car has plenty of feedback loops (not only in the cruise control module).

Systems that incorporate feedback need to work reliably even if the relationship between the desired value (e.g., the spinning speed of a motor) and the controlled value (e.g., the current applied to the motor) is not straightforward. Delay effects, inertia, and nonlinearities make life more interesting, and so do external conditions. For example, a cruise control device needs to apply more torque on the motor when the car is climbing a hill. It should keep overshoots as small as possible to avoid a speeding ticket at the top.

On the theoretical side, these problems have been well known for years and are covered by the control system theory, which explains how to design an optimized control loop for a given problem, at least if the problem is well formalized. The theory—which started with James Clerk Maxwell in 1868, followed by many other inspired mathematicians such as Alexander Lyapunov and Harry Nyquist—is not simple. The first few pages of books on control systems usually start with mathematical notions such as pole placement, Z-transforms, and sampling theorems, which nonspecialists may find

Note: This chapter is a corrected reprint of the article "The darker side: PID control without math," *Circuit Cellar*, no. 221, December 2008.

© Elsevier Inc.
DOI: 10.1016/C2009-0-20196-6

difficult to deal with, even if good books have adopted a more engineering-oriented approach (e.g., Tim Wescott's *Applied Control Theory for Embedded Systems*, Newnes, 2006). Should you give up? No. Fortunately, some classical control algorithms are applicable to many problems. More important, for design people like us, some of these algorithms are easy to implement in firmware or hardware.

In this chapter I will present the ubiquitous proportional integral derivative (PID) control, which has one interesting characteristic: It is tuned by only three or four parameters. As you will see, these parameters can be determined by empirical methods, even if the exact behavior of the controlled process is unknown. Don't get me wrong. I am not saying that a PID will solve any control problem, especially if the system has multiple inputs and outputs (such applications will be covered in the next chapter). But you may want to give PID a try before digging into more complex solutions. In any case, it's a must-have for every engineer, so let's go with PID!

A Basic Case

The most basic example of a feedback loop is a temperature controller used to drive a heater or cooler to get a precise temperature (or temperature profile) on the device's temperature sensor. In 1987, I bought a Weller VR20 soldering iron, which had a built-in resistive temperature sensor. At that time, I couldn't afford the corresponding power supply, so I built my own (see Figure 17.1).

This was easy. I used a 24-V transformer, a potentiometer to set the temperature, a digital voltmeter (Remember the old CA3161/CA3162 three-digit voltmeter chipset?), and a crude "control system." I used an LM311 comparator to drive an output TRIAC on and off, whether the measured temperature was above or below the threshold. For control system specialists, this is the most basic form of a "bang-bang" control algorithm—for obvious reasons. (Just imagine yourself driving your car with only two settings: throttle fully open or brakes fully engaged.) It worked, but the temperature regulation was oscillating at around 3°C above and below the preset value. That corresponded to an 18-mV oscillation on the sensor measurement (see Figure 17.2).

Why such a deceptive result? Simply because the heater is not in direct contact with the sensor and because the assembly does not weigh 0 g. The heat takes some time to go

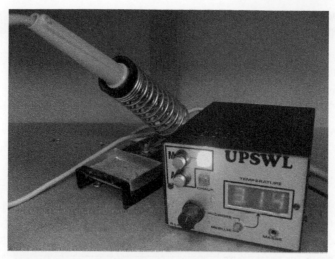

Figure 17.1: This is the power supply and controller that I built in 1987, and my Weller VR20 soldering iron. Its first basic "threshold comparator" design was working, but it had a ±3°C oscillation around the preset temperature. A simple control improvement reduced the oscillation to ±1°C.

from the heater to the tip, so if you wait for the sensor to reach the target temperature before switching the power off, it is already too late. The heat will continue to flow from the heater to the tip and you will get a significant overshoot, which ultimately causes an oscillation around the threshold. It is easy to model this behavior (see Figure 17.3).

Why not try to simulate this system under Scilab? The code starts here and continues on page 290:

```
//-------------------------------------------------------------------------
// Physical parameters
//-------------------------------------------------------------------------

// Hypothesis :
// - The heater has an homogeneous temperature, and heats the tube
// - The tube has an homogeneous temperature, heats the tip and
//   dissipates
// - The tip has an homogeneous temperature, and dissipates.
```

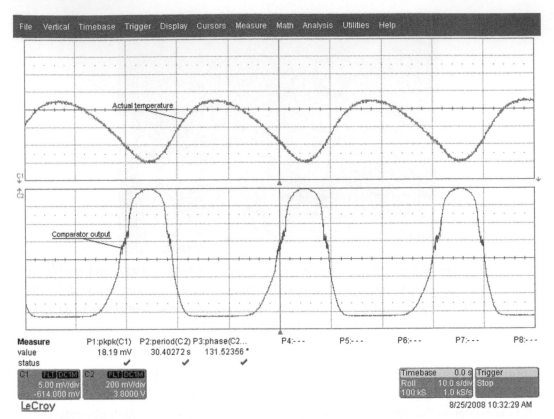

Figure 17.2: This plot was taken on my LeCroy WaveRunner 6050A oscilloscope connected to the iron controller in its initial version. The top channel is the sensor measurement. The bottom is the command sent to the output TRIAC. In this first design, I used a simple comparator. Both curves are exactly in phase; however, as a consequence the oscillation is high because of thermal inertia: 19 mV peak to peak corresponding to 6°C. Heating cycles are roughly 30 s each.

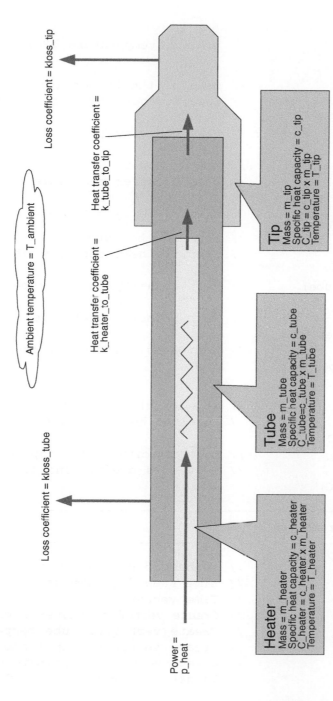

Figure 17.3: This is the thermal latency model that I used. The heater is heating a tube, which heats the iron tip. I assumed the respective temperatures of the heater, tube, and tip to be homogeneous. The heat power transferred from the heater to the tube is supposed to be a constant factor multiplied by the temperature difference between both elements. The same goes for the transfer from the tube to the tip or to ambient air. This simplistic model enabled me to get reasonably realistic simulation results.

```
// Ambiant
T_ambiant=20;               // ambiant temperature (°C)

// Heater
m_heater=20;                // mass of the heater (g)
c_heater=0.385;             // Specific heat capacity of copper
   (J per g per °C)
C_heater=m_heater*c_heater; // Heat capacity of the heater (J
  per °C)
p_heat=80;                  // Heater power (W)
k_heater_to_tube=1.0;       // Heat transfer coefficient from
   heater to tube (W per °C)

// Tube
m_tube=20;                  // mass of the tube (g)
c_tube=0.385;               // Specific heat capacity of copper
(J per g per °C)
C_tube=m_tube*c_tube;       // Heat capacity of the tube (J per
°C)
k_tube_to_tip=0.4;          // Heat transfer coefficient from
tube to tip (W per °C)
kloss_tube=0.04;            // Heat loss coefficient from tube
  to ambiant (W per °C)

// Tip
m_tip=10;                   // mass of the tip (g)
c_tip=0.385;                // Specific heat capacity of copper
   (J per g per °C)
C_tip=m_tip*c_tip;          // Heat capacity of the tip (J per °C)
kloss_tip=0.02;             // Heat loss coefficient from tip to
ambiant (W per °C)

//-------------------------------------------------------------
// Simulation parameters
//-------------------------------------------------------------

nsteps=3000;                // number of simulation steps
dt=0.2;                     // time step for simulation (s)
t=0:dt:nsteps*dt;           // Time vector
T_heater=0*t;               // Create vector of heater temperatures
T_tube=0*t;                 // Create vector of tube temperatures
T_tip=0*t;                  // Create vector of tip temperatures
Command=0*t;                // Create vector of commands (0 to 1)
T_target=350;               // Target tip temperature (°C)
```

```
//-----------------------------------------------------------------
// Simulation
//-----------------------------------------------------------------

T_heater(1)=T_ambiant;                       // Initial conditions
T_tube(1)=T_ambiant;
T_tip(1)=T_ambiant;

for i=2:length(t)                            // Main loop through time

  // ===================
  // Command calculation
  //====================

  if T_tip(i-1)<T_target then                // Simple bang-bang command
    Command(i)=1;
  else
    Command(i)=0;
  end;

  // ===================
  // Thermal simulation
  //====================

  Pgain=p_heat*Command(i-1);                 // Heater gains power if
    powered
  Pheater_to_tube = k_heater_to_tube*
(T_heater(i-1)-T_tube(i-1));                  // Heat transfer to tube
  Ptube_to_tip = k_tube_to_tip*
(T_tube(i-1)-T_tip(i-1));                     // Heat transfer to tip
  Ploss_tube = kloss_tube*
(T_tube(i-1)-T_ambiant);                      // Heat loss from tube to
  ambiant
  Ploss_tip = kloss_tip*
(T_tip(i-1)-T_ambiant);                       // Heat loss from tip to
  ambiant
  deltaT_heater=(Pgain-Pheater_to_tube)*dt/C_heater;
  T_heater(i)=T_heater(i-1)+deltaT_heater; // Heater temperature
  change
```

```
deltaT_tube=(Pheater_to_tube-Ptube_to_tip-Ploss_tube)*dt/C_tube;
T_tube(i)=T_tube(i-1)+deltaT_tube;    // Tube temperature change
deltaT_tip=(Ptube_to_tip-Ploss_tip)*dt/C_tip;
T_tip(i)=T_tip(i-1)+deltaT_tip;    // Tip temperature change


end;

// ===================
// Plot results
// ===================

subplot(3,1,1);
xtitle('Simple bang-bang control');
plot2d(t,T_heater,rect=[1,0,t(nsteps),max(T_heater)],style=[6]);
plot2d(t,T_tube,rect=[1,0,t(nsteps),max(T_heater)],style=[3]);
plot2d(t,T_tip,rect=[1,0,t(nsteps),max(T_heater)],style=[2]);
subplot(3,1,2);
plot2d(t,Command,rect=[1,-0.1,t(nsteps),1.1],style=[5]);
subplot(4,1,4);
plot2d(t(2000:3000),T_tip(2000:3000),rect=[t(2000),min(T_
tip(2000:3000)),t(3000),max(T_tip(2000:3000))],style=[2]);
halt();
xdel();
```

Figure 17.4 is the result of this simulation. Close to the behavior of my old iron controller, isn't it? As I will show you at the end of this chapter, I improved this iron controller a couple of years after its assembly and got a significantly improved regulation. On the hardware side, I do not recommend that you duplicate this design because it is based on obsolete technology. But it is a perfect example for introducing PID controls.

Proportional?

How do you improve thermal regulation? Using a full on/off drive on the heater is simple, but it is not the best solution. Why should you use 80 W of heating power if you are close to the target? The heating power should be reduced further as you approach the desired setpoint. The simplest way to do this is to calculate the heating

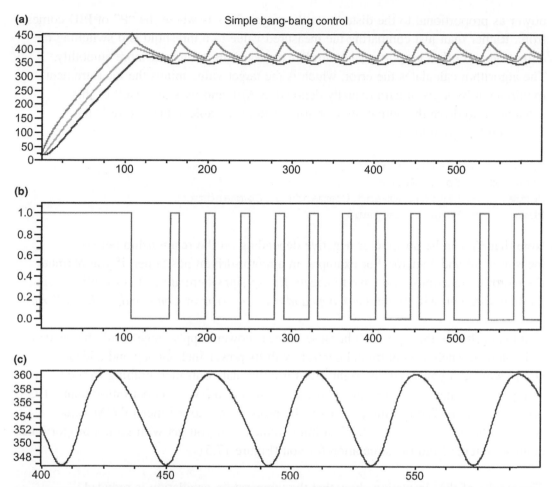

Figure 17.4: This is the simulation result for a simple "bang-bang" comparator system like the one used in the initial version of my iron controller. In (a) the curves are, respectively, heater, tube, and iron temperatures. The (b) plot is the power applied on the heater. The curve shown in (c) is a zoom on the tip temperature, showing a 12°C swing around the 350°C target with the parameters I used for this simulation.

power as proportional to the distance to the target. This is where the "P" of PID comes from. Rather than just comparing the measured value to a threshold and switching the output on and off, a proportional controller manages the switching more smoothly. The algorithm calculates the error, which is the target value minus the measurement, multiplies it by a given gain (usually denoted as *Kp*), and uses the result of this calculation to drive the output after limiting it to reasonable values. Here is the corresponding pseudocode:

```
Error = actual - target
Command = Kp . Error
Command = Limit(Command,Commandmin,Commandmax)
Pheater = Pmax . Command
```

Note that *Kp* can be positive or negative depending on the relationship between measurement and actuator. For example, in a motor-driven positioner, if you permute the motor connections, you have to negate *Kp*, so signs here are just conventions—in the literature you will find either *Command = Kp × Error* or *Command = –Kp × Error*.

You may criticize this approach because a linear power supply stage would be needed to implement such a proportional control, with its power inefficiency and added complexity, and you would be right. But for a thermal controller, nothing forbids you using the "Pheater" value to directly drive a high-speed PWM output rather than a DC power generator. The heating power will be proportional to the mean PWM value, which is the desired command. What improvement can you get with such a proportional control system? I did the simulation for you (Figure 17.5).

The results of the simulation show that the temperature oscillation is reduced from 12°C to 7°C, at least under the hypothesis of the simulation. The optimal value of *Kp* must be determined for each application because it is dependent on the system parameters and your preferences. Figure 17.6 illustrates the system's behavior with different *Kp* values. My experience tells me that it is to start with low *Kp* values and increase them up to a point where oscillations and ringing start to be a little too high.

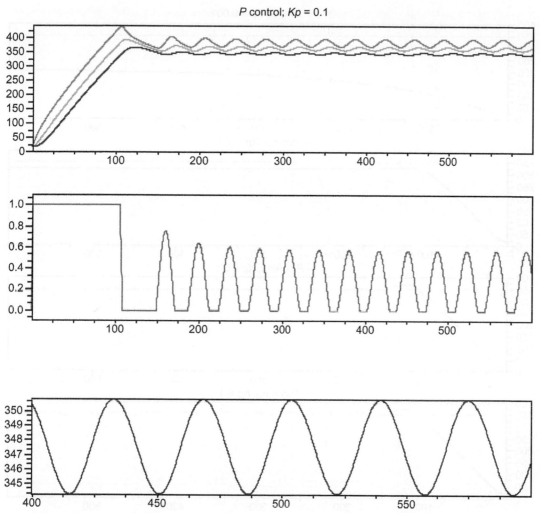

Figure 17.5: A proportional control enables you to generate an analog output value. It's low if the current measurement is close to the preset target but higher if the target is far away (medium curve). The overall temperature oscillation is reduced from 12°C to 7°C, as compared to the "bang-bang" control in Figure 17.4.

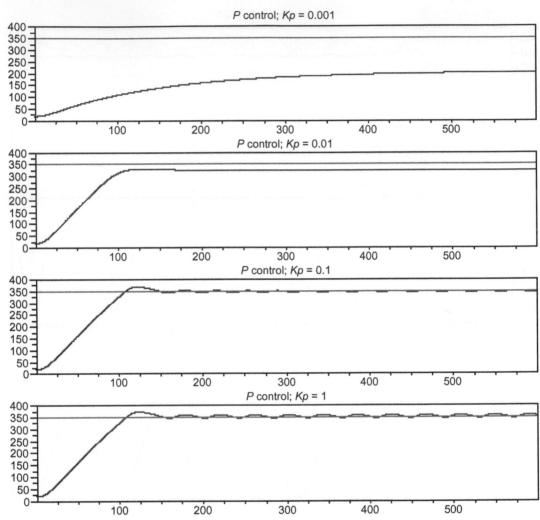

Figure 17.6: This simulation shows you the behavior of the simulated system with a proportional control and different values for the *Kp* gain. With small *Kp* values (*top*), the actual temperature takes a long time to reach the target and may not reach it. The loop is "soft." If you increase the *Kp* gain, the regulation becomes faster until an overshoot starts to appear. The optimum value is often just after the appearance of oscillations (*Kp* = 0.1 in this instance).

A Derivative Helps!

A proportional controller uses only the current measurement to determine the output value. It doesn't have memory or forecasting ability to improve the regulation. When you press the brake as you park your car in your garage, you do more than apply a pressure proportional to the distance between your car and the back wall; you also use other information, such as the speed at which your car is approaching the wall, and this makes sense. As you may remember from your youth, speed is the derivative of distance over time. A proportional-derivative (PD) control adds the derivative of the error to the calculation with another gain denoted *Kd*:

```
Error = actual - target
Command = Kp . Error + Kd . d(Error)/dt
```

For discrete time systems (e.g., microcontroller-based), the derivative of the error can be approximated as the current error minus the previous one, divided by the duration of the calculation time step, which gives the following pseudocode:

```
Previous error = Error
Error = actual - target
Command = Kp . (actual - target) + Kd/timestep . (Error -
  Previous error)
```

How does it work? If the error increases quickly, the *d(Error)/dt* term is high. Thus, for the same absolute error, the power applied on the output is higher. This enables you to return to the target quicker. In contrast, if the error decreases quickly, the *d(Error)/dt* term is negative. This reduces the power applied on the output, which reduces risks of overshoots. For simple systems, such as the one I simulated, a PD algorithm provides an impressive improvement, even if the situation is not so easy in real life (see Figure 17.7).

How do you tune the *Kd* coefficient? As illustrated in Figure 17.8, the system is more and more damped when *Kd* is higher. If *Kd* is null, we are back to a simple proportional control. An increased *Kd* reduces oscillations and overshoots, but also increases the time needed to reach a new setpoint. If you increase *Kd* even more, the system starts to be soft—often too soft. This is why I said that the *Kp* proportional parameter should usually be set a little higher than the value that gave the first oscillations because *Kd* helps reduce them. In a nutshell, with a PD regulation you can first set *Kd* and *Kp* to 0,

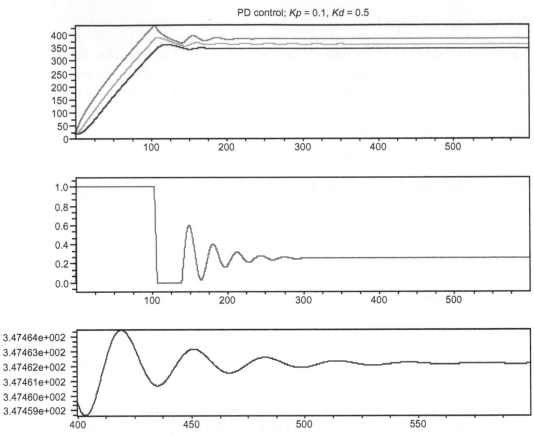

Figure 17.7: The addition of a derivative parameter gives a PD control loop. Thanks to the derivative, the oscillations are damped out and the system reaches a steady state, at least in this simulation. However, note that the stabilized value is not exactly the preset 350°C. It's a little lower (347.4°C) because of heat dissipation in ambient air.

increase Kp until small oscillations appear on the output, and then increase Kd until there is enough damping—but not more. The resulting parameter set is often close to optimal for common systems. Other authors recommend starting with a very low Kp, increasing Kd until the system stops behaving smoothly, reducing Kd a little, and then tuning Kp. I have not experimented with this method, but it makes sense too.

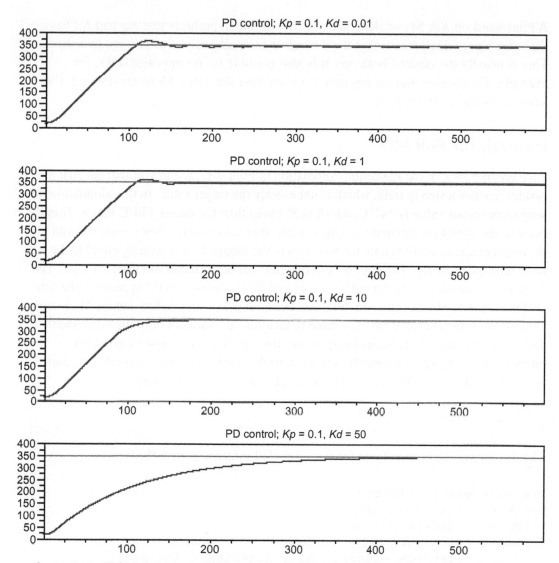

Figure 17.8: This is what usually happens when the derivative gain *Kd* is increased. The oscillations are damped more and more. At a given point, the system is no longer oscillating but starts to elongate to reach its target. The optimal setting depends on the system constraints.

A final word on *Kd*: My explanation started with the hypothesis that *Kp* and *Kd* have the same sign, which means that the derivative term allows you to prevent overshoot. This is usually the desired behavior. It is also possible to use opposite signs, for example, *Kp* positive and *Kd* negative for some specific cases. Honestly, though, I've never used this configuration.

Integral, for Full PID

Take another look at the PD control simulation in Figure 17.7, and you will see that the system reaches a steady state, which is not exactly the target value. In this simulation, the long-term sensor value is 347°C, which is 3° lower than the preset 350°C target. This is because the system is dissipative, which means that some energy flows from the iron to the ambient air. At equilibrium, the heat loss to the ambient air is exactly equal to the 3°C error multiplied by *Kp*. The system is stable, but will never reach its target setpoint: The error stays constant, so its derivative is null and *Kd* is useless. "PID" appears in the title of this chapter. I already covered "PD" feedback loops, so we need to add the "I" in order to avoid such long-term errors. You have to take into account not only the error and its derivative over time but also its integral over time. If there is a constant error, this integral will increase over time. If you add it to the command through another *Ki* gain, the equilibrium state will be forced to be exactly at the setpoint value:

```
Error = actual - target
Command = Kp . Error + Kd . d(Error)/dt + Ki . Integral (Error.dt)
```

The usual implementation form for time-sampled systems is as follows:

```
Previous error = Error
Error = actual - target
Integral = Integral + error
Command = Kp . (actual - target) + Kd/timestep . (Error -
          Previous error) + Ki . Timestep . Integral
```

This works, and it is the final form of the PID control algorithm (see Figure 17.9). However, please take care. The integral term must be manipulated with caution. Contrary to the *Kd* and *Kp* gains (at least with reasonable values), an improper *Ki* gain can make the system unstable. Moreover, the effect of *Ki* is usually opposed to *Kd*: A higher *Ki* gives a higher oscillation and a longer stabilization time.

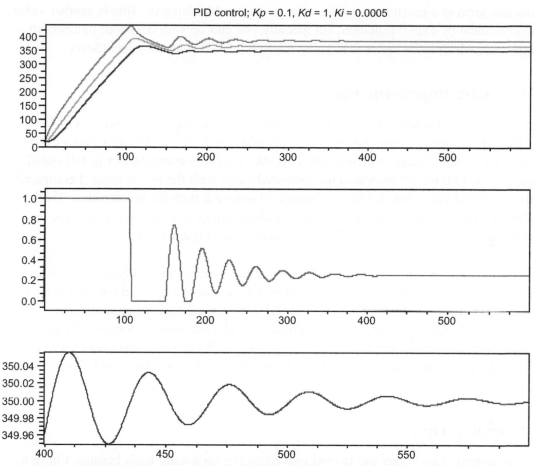

Figure 17.9: The addition of an integral term enables you to ensure that the steady point is equal to the preset 350°C target. However, note that the oscillations and stabilization time are a little higher than with a simpler PD control loop, as in Figure 17.7.

Practically speaking, it is always best to start with $Ki = 0$. If it is mandatory, increase Ki a little, after determining the optimal Kp and Kd parameters, just to a value that provides a good, long-term convergence. Then you will probably need to retune Kd and Kp to re-adapt the short-term behavior, and then Ki again, and so on. Another good way of reducing the risk of instability is to limit the maximum and minimum value of the integral term to a given range with a new parameter *MaxIntegral*. This is another value to determine by experimentation; but globally you end up with only four numbers to optimize, which is far easier than going through the full control system theory.

Hardware Improvements

It is time to go back to my dear 1987 soldering iron regulator. What did I do in 1989 to reduce the temperature oscillations? I simply added a differential term. I didn't even change the output stage, which is still a TRIAC driven by a comparator in full on/off mode. But I no longer compared the measured value with the preset value. I compared the measured value plus *Kd* times d(*measured value*)/dt with the preset value. Think about it. This is exactly the same as the PD algorithm as long as the preset value is constant. A PD control loop can be 100% analog (see Figure 17.10).

What were the actual improvements? Compare the oscillogram in Figure 17.11 with the initial one (see Figure 17.2). The addition of a *Kd* parameter reduced the temperature oscillation from 6°C to 1.3°C. That's not bad with just a couple of op-amps more. Figure 17.11 shows the derivative term in action. The output is no longer fully in phase with the sensor. It starts to increase as soon as the sensor temperature starts to reach its maximum, even if the actual temperature is still above the target. This is anticipation.

Wrapping Up

To be honest, I no longer use my old iron controller on a daily basis because I have a newer one. However, I still use it from time to time, even if it is not lead-free. Anyway, I hope that I have demonstrated that PID regulations are simple to code, or even to wire with a couple of op-amps. Moreover, they are easy to tune, at least for simple systems. Using only the proportional term may give good results. The derivative term can be added to provide damping (or shaping) of the response. The addition of an integral term ensures that there will be no systematic errors. If the average error is not zero, the integral term will increase over time, reducing the error. However, it is a little more difficult to manage than the *Kp* and *Kd* terms because it can make the system unstable. Handle it with care, or add limits to the values it allows. Don't forget that these coefficients can also be negative.

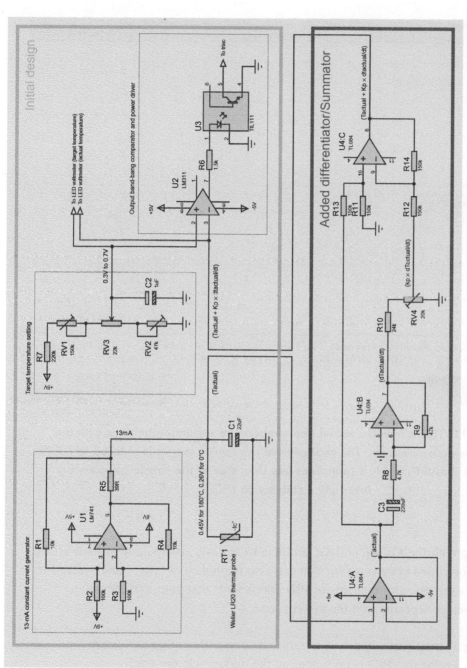

Figure 17.10: At the top is the original schematic of the temperature control section of my iron controller. A 13-mA constant current generator was driving the resistor temperature sensor, providing a voltage that is roughly proportional to the temperature. The voltage was simply compared to a preset threshold through an LM311 comparator and this comparator then drives a TRIAC through an optoisolator. The modification involved the addition of a quad TL084 op-amp to buffer (U4.A), derive over time (U4.B), and sum (U4.C) the signal with its derivative. The potentiometer RV4 sets the *Kd* gain.

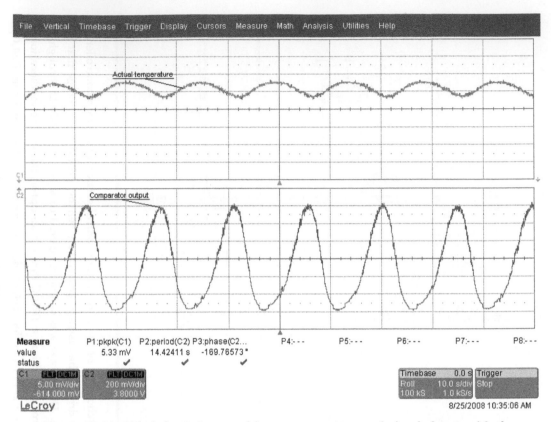

Figure 17.11: This is how the actual iron temperature regulation behaves with the addition of the derivative term. The oscillation of the sensor output is reduced to a little more than 5 mVPP, which is four times less than that of the simple comparator, providing a stability of 1°C.

Finally, playing with the *Kp*, *Ki*, and *Kd* coefficients is easy, and looking at their effects on a real-life controlled system is fun. You can also implement a "poor man's adaptive PID" with, for example, different sets of PID coefficients that your system can automatically select depending on the system state.

Consider using a small controlled system rather than experimenting at a nearby nuclear plant. Anyway, I hope that PID is no longer on the darker side for you!

CHAPTER 18

Linear Control Basics

In 1998, I developed the eZ-Stunt prototype. This project was an experimental platform built using a standard low-cost model car with an added vertical bar freely articulated on the top (Figure 18.1). Such a system is called an inverted pendulum and it is fundamentally unstable: If you leave the bar, it falls.

The game (in other words, the required embedded system) is to develop a control algorithm to automatically move the car back and forth to keep the bar as vertical as possible, even if someone gently touches it from time to time! Playing that game manually is impossible at least for more than a few hundred milliseconds, so eZ-Stunt was a very good example of the power of real-time control systems.

The on-board controller was a miniature board built around a Zilog Z80S183 mixed-signal microcontroller, which was popular in the late 1980s. Only one captor was used: a potentiometer that allowed measuring the bar angle. The microcontroller then continually updated its only controlled output: the command of the voltage applied to the main wheels motor. Even if no sensor was used to measure the position of the car, the control algorithm kept the car around its initial position, thanks to an integrated software-based position estimator. The overall eZ-Stunt hardware architecture is shown in Figure 18.2.

The basis for the eZ-Stunt platform was a standard toy-class RC car that I bought for less than $15. I even kept the original on-board radio receiver and just cut some tracks

© Elsevier Inc.
DOI: 10.1016/C2009-0-20196-6

Figure 18.1: Buy a low-cost RC-driven car, replace the radio with an on-board microcontroller, and add an unstable vertical bar. You get the eZ-Stunt, an experimental inverted pendulum platform.

on the PCB between the receiver chip and the transistor-based power amplifiers. The main modification of the car, beyond the processor board, was the addition of an articulated vertical 40-cm (15-in) aluminum extruded bar. This bar was directly fixed onto the axle of a high-precision potentiometer, fixed on a home-made aluminum stand screwed on the car top (Figure 18.3).

In this chapter I will not talk about the eZ-Stunt hardware or the actual embedded firmware since this was a one-off project that you will probably never have to duplicate.

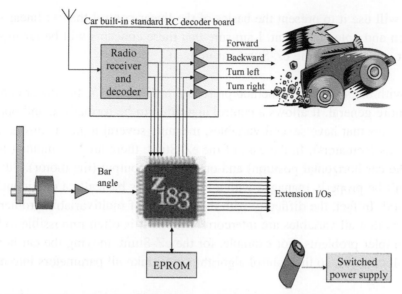

Figure 18.2: The eZ-Stunt electronic board was based on a Z80S183 processor, which was interfaced to the existing motor control board. Only the forward/backward command was used, and only one sensor was added: a potentiometer to provide a reading of the bar angle.

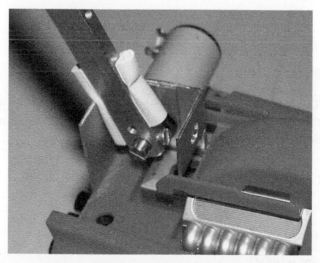

Figure 18.3: A close-up of the potentiometer axis. A 10-turn model was used to reduce friction.

However, I will use it to present the basics of the linear control theory: linear system formalization and pole placement. I am sure that these concepts will be far more useful in your next projects than the eZ-Stunt car in itself.

Compared with the simple PID method presented in Chapter 17, the linear control method is more general. It allows a control algorithm to be formalized and optimized even for systems that have several variables, meaning several inputs (sensors) and/or several outputs (actuators). In the case of the eZ-Stunt, there are two inputs (the bar angle and the car horizontal position) and one physical output (the motor). Such a problem can't be properly managed with a PID, which is basically a one input/one output method. In fact, the difficulty with the majority of multivariable problems such as eZ-Stunt is that all variables are intercorrelated, so it is often impossible to break them into simpler problems. For example, for the eZ-Stunt, moving the car horizontally inevitably tilts the bar, so the control algorithm must take all parameters into account.

Just a caution: Because of its subject, this chapter will be a little more mathematically oriented than the other chapters—and that's the reason this is the last one in this book. I will try to explain any infrequent notion, but don't hesitate to jump over the sections that may seem too theoretical and grab only the fundamental concepts.

Okay, let's go.

Physical System Modeling

In physical system modeling, the first task is to establish the mechanical model of the system to be controlled. Let's define a mechanical model for the eZ-Stunt platform. First, some definitions.

The car is moving in a straight line, and its parameters are as follows:

Car horizontal position: x (m)

Car speed: v (m/s)

Car acceleration: u (m/s^2)

The bar is articulated on the car's body, and its parameters are as follows:

Bar length: l (m)

Bar weight: M (kg)

Added mass at the end of the bar: m (kg)

Bar angle from the vertical: θ (rad)

Bar angular velocity: ω (rad/s)

The force resulting from the mass of the bar and the extra mass at its extremity is $(M + m)g$, applied at the gravity center G of the bar plus extra mass (Figure 18.4).

Now you will have to remember some of your mechanics lessons or trust me if you don't. G is the center of gravity of the oscillating system (bar and extra mass at its end). It is defined by

$$\overrightarrow{HG} = \frac{1}{M+m}\int \overrightarrow{HM}.dm = \frac{M+m/2}{M+m}l.\left(\sin(\theta).\vec{i} - \cos(\theta).\vec{j}\right)$$

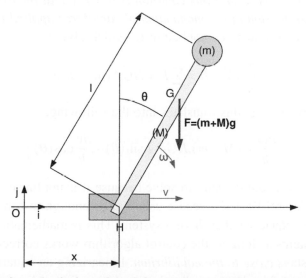

Figure 18.4: This is a physical model of the eZ-Stunt platform. It is very important for such applications to have a well-defined model that takes into account parameter-to-parameter dependencies through the underlying physics.

Similarly, the inertial momentum of the bar structure is

$$J = \int HM^2.dm = (M + m/3).l^2$$

Now we have to model the kinematic quantities. The quantity of movement of the bar is:

$$\vec{P} = \int \overrightarrow{V_M}.dM = (M + m).\vec{V_G} = (M + m).\omega.\vec{k} \wedge \overrightarrow{HG} + (M + m).\vec{V}$$

The kinematic momentum of the bar calculated in H is

$$\vec{\sigma} = \int \overrightarrow{HM} \wedge \overrightarrow{V_M}.dm = (m + M).\overrightarrow{HG} \wedge \vec{V} + J\omega.\vec{k}$$

The dynamic momentum of the bar is

$$\vec{\Phi} = \frac{d\vec{\sigma}}{dt} + \vec{V} \wedge \vec{P}$$

With all these calculations, we have enough material to write an equation giving a "behavioral" model of the system—meaning one that expresses the relationship between bar angle, car position, speed, and acceleration. This is called the dynamic equation of the system. *The easiest way to find this equation is to write that the dynamic momentum of the bar is equal to the sum of the momentum of each force applied to the system* (i.e., the bar), here only the weight force applied in G, which gives

$$\frac{d\vec{\sigma}}{dt} + \vec{V} \wedge \vec{P} = \sum \vec{F} = \overrightarrow{HG} \wedge (M + m)\vec{g}$$

Projected on the vertical axis, this translates into the following:

$$J.\frac{d\omega}{dt} = (M + m).HG.\left(g.\sin(\theta) - \frac{dv}{dt}\cos(\theta) \right)$$

The last equation is exact but difficult to solve because it is nonlinear. The title of this chapter is "Linear Control Basics" because usually it is enough to consider a linear approximation of the equations that drive a system. This is mathematically true for the vast majority of systems as long as the control algorithm works correctly, meaning as long as the system stays *close to the equilibrium*. Linearizing an equation means considering that all variables have *small values*, so using first-order approximations makes sense. For example, sin(x) can be approximated as x, and cos(x) can be approximated as 1. Let's be honest: It doesn't work all the time—some complex control problems can't be linearized—or it provides deceptive results. Nonlinear control algorithms do exist but are far more complex. Let's stay with the simple linear control theory in this book, which is largely adequate for the eZ-Stunt.

By linearization of the preceding equation, and after dividing each side by J, we find

$$\frac{d\omega}{dt} = \frac{(M+m).HG}{J}.\left(g.\theta - \frac{dv}{dt}\right) = \frac{(M+m).HG}{J}.(g.\theta - u)$$

This can be rewritten in simplified form:

$$\frac{d\omega}{dt} = a.q + b.u$$

with

$$\begin{cases} a = \dfrac{(M+m).HG}{J}g = \dfrac{M+\dfrac{m}{2}}{M+\dfrac{m}{3}}.\dfrac{g}{l} \\[4mm] b = -\dfrac{a}{g} \end{cases}$$

The previous equation is the linear model of the system, and it really is a simple equation as a and b are just numeric coefficients fixed by the size and weight of the system. If you look closely, you will find that this equation is in fact quite clear: It says that the angular acceleration of the bar is proportional to its angle (if the bar is tilted, it will fall) and proportional to the acceleration of the car.

Linear Control Standard Format

Were we mathematically speaking? We found a first-order equation giving the evolution of bar acceleration based on the acceleration of the car (the command u) and the bar's previous angle. The linear control theory has some fundamental theorems that enable us to quickly check if the system can be effectively controlled with a linear feedback loop, but first we need to write the previous equation in the following standardized matrix form:

$$\begin{cases} \dfrac{dX}{dt} = A.X + B.u \\ \quad Y = C.X \end{cases}$$

The first line should look like the previous equation in bold, but it is in a much more general form. X is the system's state vector. Basically, the equation needs to include all system variables: positions and speeds usually, which can be either visible to the external world or internal to the system. U is the command, Y is the observable vector and is the set of "visible" parameters, which can be a subset of X or derived values. A, B, and C are fixed matrices.

In the case of eZ-Stunt, the system state vector X is constructed with the car position (x), the car speed (v), the bar angle (θ), and the bar angular velocity (ω).

Now we have to do some calculations, but I love to use tools to make my life easier. Here we need to manipulate symbolic expressions and matrices to solve linear systems and so on. Let's use the free Maxima algebra software and its accompanying WsMaxima user interface (Figure 18.5), which are very well suited to the job. Don't

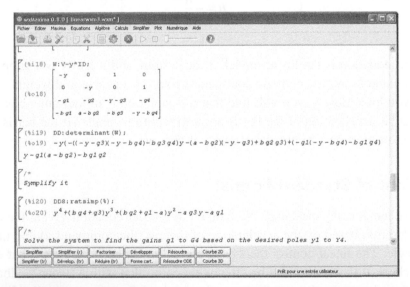

Figure 18.5: Maxima is free algebra software available from SourceForge under either the Linux or the Windows operating system.

be confused. Maxima is not a numerical calculation tool like Scilab but a symbolic tool. It manages equations and expressions even if the numerical values are not known.

First, we need to define the system state vector X and the system output vector Y. The state vector includes position x, angle θ, and the corresponding speeds v and ω. The output vector is only x and θ:

```
(%i1)  X: matrix([x],[theta],[v],[omega]);
                          [   x     ]
                          [         ]
                          [ theta   ]
(%o1)                     [         ]
                          [   v     ]
                          [         ]
                          [ omega   ]

(%i2)  Y: matrix([x],[theta]);
                          [   x     ]
(%o2)                     [         ]
                          [ theta   ]
```

Then we define the system dynamic matrix A and observation matrix C. Remember that with these matrices the system dynamic must be defined as $dX/dt = AX + Bu$, and the output vector as $U = CX$. This is the standard form of a linear system:

```
(%i3)  A: matrix([0,0,1,0],[0,0,0,1],[0,0,0,0],[0,a,0,0]);
                        [ 0  0  1  0 ]
                        [            ]
                        [ 0  0  0  1 ]
(%o3)                   [            ]
                        [ 0  0  0  0 ]
                        [            ]
                        [ 0  a  0  0 ]
(%i4)  B: matrix([0],[0],[1],[b]);
                            [  0  ]
                            [     ]
                            [  0  ]
(%o4)                       [     ]
                            [  1  ]
                            [     ]
                            [  b  ]
```

You may think that this is a complex way of expressing simple things, but with these notations the expression *dX/dt = AX + Bu* is equivalent to the linear model of the system. Let's ask Maxima to calculate *AX + Bu*:

```
(%i14)  A.X+B*u;
                              [       v       ]
                              [               ]
                              [     omega     ]
(%o14)                        [               ]
                              [       u       ]
                              [               ]
                              [ b u + a theta ]
```

So *dX/dt = AX + Bu* means that

$d(x)/dt = v$ (the definition of *v*)

$d(\theta)/dt = \omega$ (the definition of *ω*)

$d(v)/dt = u$ (the definition of *u*)

$d(\theta)/dt = a\ \theta + bu$ (the linear equation of the system)

With this formalism we then have a compact and linear representation of the system. More important, it will allow us to use the results of the linear control theory.

Observable and Reachable?

Two preliminary checks are now needed. First, we need to check if the system is *observable*, that is, can the system state be properly monitored from the output vector? Mathematicians have proven that it can if the rank of the vector {*C, CA, CAA, CAAA*} is equal to 4. Let's calculate this matrix with Maxima:

```
(%i6)  C;
                              [ 1  0  0  0 ]
(%o6)                         [            ]
                              [ 0  1  0  0 ]

(%i7)  C.A;
                              [ 0  0  1  0 ]
```

```
(%o7)                              [              ]
                                   [ 0  0  0  1 ]
(%i8)  C.A.A;
                                   [ 0  0  0  0 ]
(%o8)                              [              ]
                                   [ 0  a  0  0 ]
(%i9)  C.A.A.A;
                                   [ 0  0  0  0 ]
(%o9)                              [              ]
                                   [ 0  0  0  a ]
```

No problem. The rank is always 4 because none of these matrices can be null. Now we need to check if the system is *reachable*—that is, if there is any way to control the system and keep the bar vertical. A linear control system is reachable if the rank of the matrix {*B, AB, AAB, AAAB*} is equal to 4. Let's see:

```
(%i10)  B;
                                   [  0  ]
                                   [      ]
                                   [  0  ]
(%o10)                             [      ]
                                   [  1  ]
                                   [      ]
                                   [  b  ]
(%i11)  A.B;
                                   [  1  ]
                                   [      ]
                                   [  b  ]
(%o11)                             [      ]
                                   [  0  ]
                                   [      ]
                                   [  0  ]
(%i12)  A.A.B;
                                   [  0  ]
                                   [      ]
                                   [  0  ]
(%o12)                             [      ]
                                   [  0  ]
                                   [      ]
                                   [ a  b ]
```

```
(%i13)  A.A.A.B;
```

$$
(\%o13) \qquad
\begin{bmatrix}
0 \\
\\
a \ b \\
\\
0 \\
\\
0
\end{bmatrix}
$$

The result is that the system is indeed observable and reachable—two complicated words to say that the eZ-Stunt game is really possible and that this bar can be really kept vertical with a linear feedback. In other words, it is possible to find four gains, $g1$, $g2$, $g3$, and $g4$, which respectively multiplied by the car position, car speed, bar angle, and bar angular velocity give the required acceleration to apply to the car:

$$U = -(g1 \times x + g2 \times v + g3 \times q + g4 \times \omega)$$

Let's define this gain vector and the associated command under Maxima:

```
(%i14)  G:matrix([g1,g2,g3,g4]);
(%o14)                    [ g1  g2  g3  g4 ]

(%i15)  u:-G.X;
(%o15)           - g1 x - g3 v - g2 theta - g4 omega
```

Pole Placement!

How do you calculate the correct values for $g1$, $g2$, $g3$, and $g4$? This is the only difficult job! Several theoretical methods exist, but here I will present the quite simple pole placement method. With it, the user still has to fix four values (the poles, which define the dynamic behavior of the controlled system), but the four gains are then automatically calculated from the four poles. More important, with this method you can be sure that the system will be stable as long as these four poles are negatives. Moreover, the regulation will be more and more straight as the poles are far from zero.

Mathematically, the poles of the system are calculated as eigenvalues of the matrix $A - BG$, that is, as the roots of the following equation:

$$\det((A - BG) - \lambda.I) = 0$$

Let's calculate this determinant with Maxima (I used the variable *y* in place of lambda to simplify the writing):

```
(%i16)  V: A-B.G;
                    [        0        0        1        0        ]
                    [                                           ]
                    [        0        0        0        1        ]
(%o16)              [                                           ]
                    [      - g1     - g2     - g3     - g4      ]
                    [                                           ]
                    [  - b g1  a - b g2   - b g3   - b g4       ]

(%i17)  ID:ident(4);
                    [ 1  0  0  0 ]
                    [            ]
                    [ 0  1  0  0 ]
(%o17)              [            ]
                    [ 0  0  1  0 ]
                    [            ]
                    [ 0  0  0  1 ]

(%i18)  W:V-y*ID;
                    [      - y        0        1        0        ]
                    [                                           ]
                    [       0       - y        0        1        ]
(%o18)              [                                           ]
                    [     - g1      - g2    - y - g3   - g4      ]
                    [                                           ]
                    [ - b g1   a - b g2    - b g3   - y - b g4  ]

(%i19)  DD: determinant(W);
(%o19) - y (- ((- y - g3) (- y - b g4) - b g3 g4) y - (a - b
          g2) (- y - g3) + b g2 g3) + (- g1 (- y - b g4) - b
          g1 g4) y - g1 (a - b g2) - b g1 g2

(%i20)  DDS: ratsimp(%);
              4             3                  2
(%o20)    y  + (b g4 + g3) y  + (b g2 + g1 - a) y  - a g3 y - a g1
```

The previous equation is a polynomial of degree four. Such an equation has usually four roots, meaning four values of y, which satisfy this equation. These four values are the poles of the system, and we want to fix their values to whatever pole values the user will want (e.g., $l1$, $l2$, $l3$, and $l4$). Remember that all of these values must be negative for a stable system. If we write that the equation satisfies each of the four poles $l1$ to $l4$, then we will have four equations, with four unknown variables: the gains $g1$, $g2$, $g3$, and $g4$. Look at the equations again and you will find that each equation is a linear combination of $g1$ to $g4$, so all you have to solve is a set of four linear equations of four variables. An easy task with Maxima but a very boring one manually.

Let's first duplicate the equations with the four different variable names, and then solve the system to find $g1$, $g2$, $g3$, and $g4$.

```
(%i21)  E1:  (ratsubst(y1,y,DDS)=0);
        4        3          3       2            2       2
(%o21)  y1  + g3 (y1  - a y1) + b g4 y1  + g1 (y1  - a) + b g2
  y1  - a y1 = 0

(%i22)  E2:  (ratsubst(y2,y,DDS)=0);
        4        3          3       2            2       2
(%o22)  y2  + g3 (y2  - a y2) + b g4 y2  + g1 (y2  - a) + b g2
  y2  - a y2 = 0

(%i23)  E3:  (ratsubst(y3,y,DDS)=0);
        4        3          3       2            2       2
(%o23)  y3  + g3 (y3  - a y3) + b g4 y3  + g1 (y3  - a) + b g2
  y3  - a y3 = 0

(%i24)  E4:  (ratsubst(y4,y,DDS)=0);
        4        3          3       2          2       2
(%o24)  y4  + g3 (y4  - a y4) + b g4 y4  + g1 (y4  - a) + b g2
  y4  - a y4 = 0

(%i25)  linsolve([E1,E2,E3,E4],[g1,g2,g3,g4]);
                    y1 y2 y3 y4
(%o25)  [g1 = - ──────────────── ,  g2 =
                       a
```

$$y1 \ (a \ (y4 + y3 + y2) + y2 \ y3 \ y4) + a \ (y2 \ (y4 + y3) + y3 \ y4) + a^2$$
─── ,
$$a \ b$$

$$g3 = \frac{y1\ (y2\ (y4 + y3) + y3\ y4) + y2\ y3\ y4}{a},$$

$$g4 = -\frac{y1\ (y2\ (y4 + y3) + y3\ y4 + a) + a\ (y4 + y3 + y2) + y2\ y3\ y4}{a\ b}]$$

```
(%i26) factor(trigreduce(%));
```

$$(\%o26)\ [g1 = -\frac{y1\ y2\ y3\ y4}{a},$$

$$g2 = \frac{y1\ y2\ y3\ y4 + a\ y3\ y4 + a\ y2\ y4 + a\ y1\ y4 + a\ y2\ y3 + a\ y1\ y3 + a\ y1\ y2 + a^2}{a\ b},$$

$$g3 = \frac{y2\ y3\ y4 + y1\ y3\ y4 + y1\ y2\ y4 + y1\ y2\ y3}{a},$$

$$g4 = -\frac{y2\ y3\ y4 + y1\ y3\ y4 + y1\ y2\ y4 + a\ y4 + y1\ y2\ y3 + a\ y3 + a\ y2 + a\ y1}{a\ b}]$$

Just fix the $l1$, $l2$, $l3$, and $l4$ pole values to arbitrary negative numbers, calculate the gains $g1 \dots g4$ with the previous equations, and use them in a linear feedback loop based on the equation $U = -(g1 \times x + g2 \times v + g3 \times \theta + g4 \times \omega)$ to calculate at each time step the required acceleration of the car.

The linear control theory allows us to claim that any set of four negative pole values will give a stable control loop, but only in an ideal world without any mechanical limitations (possibly nearly infinite acceleration in particular). Several theoretical methods exist to set the four pole values, but here a simple trial and error method can be used with some heuristic rules:

- The four pole values should be "spread" (i.e., not too close to each other).
- The four values should be between −1 and around −20.
- The highest negative values make control more difficult.

Here experience is a must!

Wrapping Up

I wrote the small eZ-Stunt simulation software on my PC. This software implements all the real-time calculation and the simulation code to mimic the evolution of the real mechanical system, adding some noise in all calculations to check if the "real" system would work. This simulation code first helped to debug the real-time calculation code (that was directly reused for the embedded software) and then helped to quickly find good poles values: the best working values with higher and higher noise injected in the calculations. You will find it on the companion website; don't hesitate to play with it.*

The learning curve to understand linear systems theory may seem difficult, in particular because several mathematical concepts are required. However, the difficulties are mainly in the system-modeling phase. If you succeed in this phase, you will have all the horsepower of the theory. Moreover, the real-time calculations are then minimal: only a couple of multiplications and additions. I hope that I have convinced you that algebra tools such as Maxima are definitely a very good way to come out of the darker side!

---

*This simulation code has a relatively crude graphical user interface (written in half an hour with an old Turbo-C compiler using DOS). If you want to look at a controlled bar, just copy the files simstunt.exe and egavga.bgi into a DOS directory such as *c:\test*; launch a DOS shell; and execute simstunt.exe. Press j or k to simulate a manual action on the bar. Don't try to go off the screen–you will succeed.

Scilab Tutorial

Scilab?

I used Scilab quite heavily throughout this book for numerical calculations and examples. In this appendix you will find a brief Scilab tutorial.

Some history first. Scilab is an open-source scientific software package developed in the early 1990s by researchers from two French organizations—INRIA (National Institute for Research in Computer Science and Control) and ENPC (Ecole Nationale des Ponts et Chaussées)—which followed developments undertaken in the 1980s. The first official release, Scilab 1.1, was in January 1994.

The Scilab project has been managed since 2003 by the Scilab Consortium (see *http:// www.scilab.org*). The members of this group are still mainly French educational, research, and industrial companies: INRIA, CEA, CNES, EADS, EDF, RENAULT, Ecole Polytechnique, and ECP, among others, but Scilab now has thousands of users worldwide. The full tool suite is still open source and free, thanks to members' contributions and public funds through the Digiteo research organization. The package is not licensed under GPL but can be used for any academic or commercial application and can even be redistributed freely, with the inclusion of copyright notice as the only restriction.

What is Scilab for? It is a numerical calculation tool and development environment that will help you to work on actual numbers, not just symbolical expressions. It is functionally very close to MATLAB (© the Mathworks) but not fully compatible

© Elsevier Inc.
DOI: 10.1016/C2009-0-20196-6

because of syntactical variants. With Scilab you get plenty of toolboxes—from 2D/3D graphics to linear algebra, sparse matrices, polynomials, multiple interpolation methods, signal processing, and statistics, to name a few.

Moreover, many user contributions enhance it with pretty user interface development toolkits, TCP/IP interfaces, parallel computing support, and databases. There is a graphical modeling tool, Scicos, as well. Scilab is also a full-featured interpreted programming language.

On the computing side, at the time of this writing Scilab works on most UNIX/Linux platforms and of course on Microsoft Windows systems. You can easily link Scilab with software coded in C/C++, Java, and maybe Fortran, as well as TCL/TK.

Just go to *http://www.scilab.org*, download and install the latest version on your computer, and follow me with a couple of examples. My intent is not to transform you into a Scilab master but just to show you some of the possibilities of this tool. Note that these examples were tested on Scilab 5.1.1.

Starting Scilab

After installation, launch the Scilab application from your desktop menu. You will get a console window similar to Figure A.1. Enter any operation—here a very complex 1 + 1—and the Scilab interpreter will give you the answer.

You will soon be bored with entering commands manually, so write a Scilab script, which is nothing more than a text file, and execute it. Just open the "Applications" menu and select "Editor" to get the Scilab editor up and running. Type any Scilab code in the Editor window, save it if you wish, and execute it with the "Execute/Load all into Scilab" menu item or with the Ctrl-L shortcut (see Figure A.2).

Finally, Scilab includes very complete online help, so don't hesitate to click on the help menu in the Scilab console window or use the help command (example: "help plot").

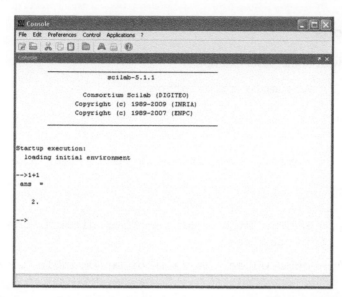

Figure A.1: The Scilab desktop is a text-mode console window.

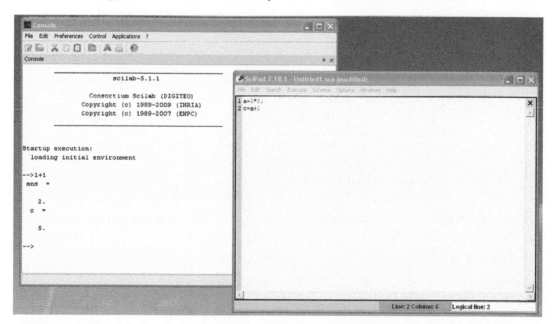

Figure A.2: The Scilab editor is just a convenient way of editing Scilab scripts. To execute scripts, use the Ctrl-L shortcut.

Basic Functions

Okay, let's start with some basic examples:

```
-->1+2          // A simple calculation
 ans =

   3.

-->a=3+4        // here the result is stored in variable a
 a =

   7.

-->c=sin(2);  // Ending with a semi-column disable the display
              // of the result

-->c            // Just enter the name of a variable to display it
 c =

   0.9092974

-->log(%pi)  // Constants starts with a % (ex : %pi, %e, etc)
 ans =

   1.1447299

-->whos -type constant  // List the content of the workspace
Name        Type         Size          Bytes
c           constant     1 by 1        24
a           constant     1 by 1        24
%inf        constant     1 by 1        24
%eps        constant     1 by 1        24
%io         constant     1 by 2        32
%i          constant     1 by 1        32

-->c=2+1.3*%i     // Complex numbers are of course supported
 c =

   2. + 1.3i

-->c^2            // Square of a complex
 ans =

   2.31 + 5.2i

-->log(c)         // Complex operations
 ans =
```

```
     0.8693551 + 0.5763752i
-->rand('normal')  // Initialise the random generator in normal mode
-->rand()          // Get a random value
 ans  =

    0.6755537
-->x=poly(0,'x')   // Seed a polynomial built on variable x
 x  =

    x
-->p=x^2-3*x-4     // Create a polynomial
 p  =
             2
  - 4 - 3x + x
-->z=roots(p)      // Calculate the roots of the polynomial
 z  =

  - 1.
    4.
-->derivat(p)      // Or derivate it...
 ans  =

  - 3 + 2x
```

Vectors and Matrices

Scilab is well suited to matrix calculations, which are usually quite boring if done manually. Let's see some examples of the syntax:

```
-->A=[1 2 3;9 5 3; 0 1 3]  // Create a 3x3 matrix
 A  =

    1.    2.    3.
    9.    5.    3.
    0.    1.    3.
-->x=[3;5;-2]              // Create a vertical vector
 x  =

    3.
    5.
  - 2.
```

```
-->h=[5 6 7]                 // Create a horizontal vector
 h =

   5.    6.    7.

-->A*x                       // Calculate a matrix by vector
product
 ans =

   7.
   46.
 - 1.

-->A*A                       // Square a matrix
 ans =

   19.    15.    18.
   54.    46.    51.
   9.     8.     12.

-->inv(A)                    // Invert a matrix
 ans =

 - 0.8    0.2        0.6
   1.8  - 0.2      - 1.6
 - 0.6    0.0666667   0.8666667

-->sin(x)                    // All operations can be applied on matrix
 ans =                       // (term by term)

   0.1411200
 - 0.9589243
 - 0.9092974

-->A(2,3)                    // Extract an element from a matrix
 ans =

   3.

-->A(2,:)                    // Extract a line from a matrix
 ans =

   9.    5.    3.
-->A(2,2:$)                  // Extract part of a line from a matrix
 ans =

   5.    3.
```

```
-->A'                   // transpose the matrix A
 ans  =

    1.    9.    0.
    2.    5.    1.
    3.    3.    3.
-->det(A)               // Calculate the determinant of A
 ans  =

  - 15.

-->y=A\x                // Solve the linear system x=A*y

 y  =

  - 2.6
    7.6
  - 3.2
-->ones(3,3)            // Generate specific matrix
 ans  =

    1.    1.    1.
    1.    1.    1.
    1.    1.    1.
-->A.*A                 // Syntax for term by term operation
 ans  =

    1.    4.    9.
   81.   25.    9.
    0.    1.    9.
```

Vectors are heavily used for iterative loops, such as generation of sampled signals or input data to systems or equations. The syntax is straightforward:

```
-->w=0:10               // Generate a vector from 0 to 10
 w  =

    0.    1.    2.    3.    4.    5.    6.    7.    8.    9.    10.
-->w=1:0.01:1.05        // Or from 1 to 1.05 with a step of 0.01
 w  =

    1.    1.01    1.02    1.03    1.04    1.05
-->w+1                  // Do some calculations on this vector
 ans  =

    2.    2.01    2.02    2.03    2.04    2.05
```

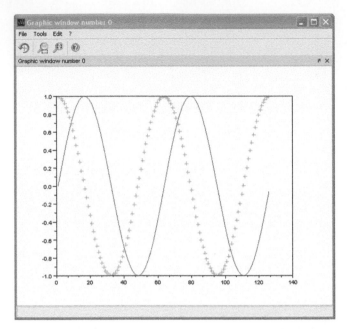

Figure A.3: Both curves are shown on the same diagram, with their respective styles.

Plotting

As illustrated in this book, Scilab includes an impressive library of graphing and plotting functions. The following are some basic examples.

```
-->x=0:0.1:4*%pi;        // Generates a vector
-->plot(sin(x),'red')    // Plot a sine in red color
-->plot(cos(x),'cya+')   // Plot a cosine in cyan, using cross dots
```

The result is shown in Figure A.3. The following code demonstrates x/y charts and 3D surfaces. See the result of this in Figures A.4 and A.5, respectively.

```
-->x=[0.5 0.2 0.6 0.9 0.45];     // Generates an x vector
-->y=[1.2 0.4 0.2 0.9 1.1];      // and an y vector
-->plot2d(x,y,style=-1,rect=[0 0 1 1.5])  // Plot their x/y chart
                                 // using dots with predefined axis
```

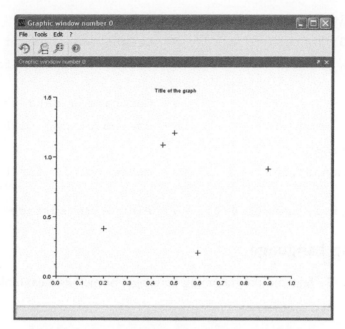

Figure A.4: X/Y scatter plots can also be plotted.

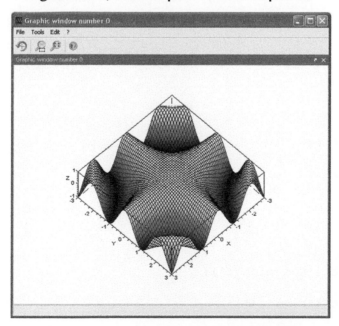

Figure A.5: An example of a 3D surface plot.

```
-->xtitle("Title of graph")      // Add a title
-->x=-3:0.1:3;                   // Create an x vector
-->M=x'*x;                       // Create a bidimentionnal
vector
-->z=cos(M);                     // Generate a data set
-->C = hotcolormap(32);          // use color map with 32 values
-->xset("colormap",C)
-->xset("hidden3d",30)           // define colors for hidden
surfaces
-->plot3d1(x,x,z, flag=[1 4 4])  // Plot a 3D surface
```

Programming Language

Scilab is also a fully featured programming language. Some basic examples:

```
function [r]=myfunct(x,y)
    if(x<y) then
        r=x;
    else
        r=y;
    end;
endfunction;

-->myfunct(6,5)
 ans  =

    5.
function guess(x)
    v=0;
    while(v<>x)
      v=input("your choice ?");
      if v<x then
        printf("too low\n");
      else
        if v>x then
          printf("too high\n");
        else
          printf("bingo !");
        end;
```

```
      end;
    end;
endfunction;

-->guess(10)
your choice ?2
too low
your choice ?11
too high
your choice ?10
bingo !

--> for i=1:5
        printf("i=%d ",i)
    end;

i=1  i=2  i=3  i=4  i=5
```

Toolboxes

Finally, in addition to the hundreds of built-in functions, you will find on *http://www. scilab.org* a long list of user contributions. Don't forget to browse them before coding something that probably already exists. Figure A.6 shows the toolboxes currently available to download.

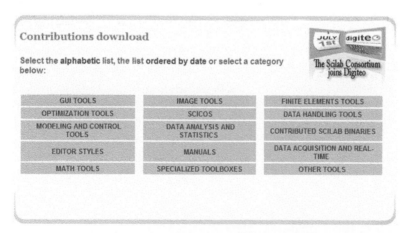

Figure A.6: A snapshot of the user-contributed toolbox collections that are available at Scilab.org at time of writing.

Complex Numbers 101

In this book I used complex numbers from time to time, particularly when discussing impedance matching. Complex numbers are a very powerful and common mathematical representation, but you may not be used to them. Here is a small introduction.

Geometrical Interpretation

Let's suppose that you have two numbers, a and b, but you want to manage them as a single entity for convenience. You can consider a and b as the coordinates of a point Z on a 2D plane. This is called the "complex plane," but it is actually a very simple construction: Just measure a distance a on the horizontal axis (called the "real axis" in complex arithmetic) and a distance b on the vertical axis (the "imaginary axis"), and you have located point Z by its rectangular coordinates (see Figure B.1).

Point Z is represented by a so-called complex number z, which is noted as follows:

$$z = a + ib$$

Sometimes the form $z = a + jb$ is used, mainly in electrical engineering, where i can be erroneously understood as a current. However, the official form is $a + ib$. This is simply special notation to differentiate the first number related to the horizontal axis and the second number related to the vertical axis. i also has a very special property: Its square is negative or, to be more exact, -1:

$$i^2 = -1$$

© Elsevier Inc.
DOI: 10.1016/C2009-0-20196-6

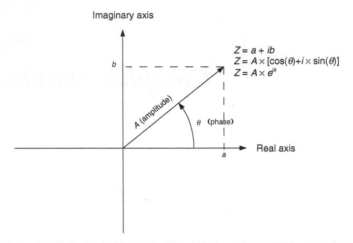

Figure B.1: Complex numbers have two coordinates (*a* and *b*), and they are usually represented as *a* + *ib* or *a* + *jb*. They can also be represented by their polar form A and *θ*.

The square of any real number is positive, so *i* is actually a very special symbol, sometimes called the square root of −1 even if this is mathematically inexact. Just consider it as convention for the moment.

Finally, some vocabulary: The complex number created by keeping the same real value while negating the imaginary part is called the complex conjugate of the number and is denoted as follows:

$$\overline{a+ib} = a - ib$$

Polar Form

Because *Z* is a point on a plane, you can also locate it by its polar coordinates—namely, the distance between the origin *O* and the point *Z* (called the "modulus" of the number *z*) and the angle between the real axis and the *OZ* line (called the "argument," or phase, of *z*).

I'm sure you remember the basic trigonometric identities:

$$a = A\cos(\phi)$$

$$b = A\sin(\phi)$$

and

$$A = \sqrt{(a^2 + b^2)}$$

$$\phi = \arctan(b/a)$$

(with the proper correction depending on the quadrant, $0°$ to $90°$, $90°$ to $180°$, etc.) The polar form of a complex number is then

$$Z = a + ib = A(\cos(\phi) + i\sin(\phi)) = Ae^{i\phi}$$

Just consider this last form as a convention, but it is actually a theorem (the Euler theorem).

Operations

Let's add two complex numbers:

$$(a + ib) + (c + id) = (a + c) + i(b + d)$$

The addition of the complex numbers is achieved by adding, respectively, their real and imaginary parts—easy.

Now let's multiply two complex numbers represented by their rectangular coordinates:

$$(a + ib) \times (c + id) = ac + iad + ibc + i^2bd = ac + iad + ibc - bd$$
$$= (ac - bd) + i(ad + bc)$$

Not as easy, is it? But what would the formula be if we used polar forms? Let's suppose that

$$z_1 = A(\cos\phi + i\sin\phi) \quad \text{and} \quad z_2 = B(\cos\alpha + i\sin\alpha)$$

and we want to know, in polar form, the product of

$$z_1 \times z_2$$

Let's calculate it. We have

$$z_1 z_2 = A(\cos\phi + i\sin\phi) \times B(\cos\alpha + i\sin\alpha)$$

Thus

$$z_1 z_2 = AB(\cos\phi + i\sin\phi)(\cos\alpha + i\sin\alpha)$$

Distributing the product and using the fact that $i^2 = -1$, we get

$$z_1 z_2 = AB[(\cos\phi\cos\alpha - \sin\phi\sin\alpha) + i(\cos\phi\sin\alpha + \sin\phi\cos\alpha)]$$

Finally, thanks to the basic trigonometric formulae, this can be simplified as

$$z_1 z_2 = AB[\cos(\phi + \alpha) + i\sin(\phi + \alpha)]$$

As we see, the multiplication of two complex numbers is quite complex when using the rectangular format but very easy in polar form: The product of two complex numbers is calculated by multiplying their moduli (amplitudes) and adding their arguments (phases). This is a fundamental property of the behavior of complex numbers. Consider for example the specific case of i, which has a modulus of 1 and an argument of 90°. Its square ($i \times i$) thus has a modulus of 1×1, which is 1, and a phase of $2 \times 90° = 180°$, so $i^2 = -1$ as expected.

In fact, the complex number theory allows us to embed the trigonometric formulae into a convenient formalism and to ease the calculations. Thus, it is not surprising that the theory is so widely used for impedance calculations since a single number can represent both the amplitude and the phase shift of an electrical signal.

Resources

Chapter 1

INRIA and the Scilab Consortium. Scilab software; available at <www.scilab.org>.
SourceForge. Quite Universal Circuit Simulator (QUCS); available at <qucs.sourceforge.net>.
Wetherell, J., 1997. Online Impedance Matching Network Designer; available at <home. sandiego.edu/~ekim/e194rfs01/jwmatcher/matcher2.html>.

Chapter 2

Agilent Technologies, Inc. AppCAD; available at <www.hp.woodshot.com>.
Brooks, D., 2000. Embedded Microstrip: Impedance Formula. Printed Circuit Design.
California Institute of Technology RF and Microwave Group. PUFF Microstrip Layout and Simulator; available at <www.its.caltech.edu/~mmic/puff.html>.
Hewlett-Packard. HP8620C Microwave sweeper, HP8754A/H26 vector network analyzer, and HP8755 scalar analyzer; available at <www.hp.com>.
Lee, T., 2004. Planar Microwave Engineering: A Practical Guide to Theory, Measurement, and Circuits. Cambridge University Press.
QUCS Team. QUCS Project; available at <qucs.sourceforge.net>.
Rogers Corp. RO4003 Substrate; available at <www.rogerscorp.com>.
Sonnet Software, Inc. Sonnet Lite; available at <www.sonnetsoftware.com/products/lite>.
The University of York Department of Computer Science. LC Filter Design Calculator; available at <www-users.cs.york.ac.uk/~fisher/lcfilter>.

© Elsevier Inc.
DOI: 10.1016/C2009-0-20196-6

Chapter 3

Engdahl, T., 2007. Time Domain Reflectometer; available at <www.epanorama.net/circuits/tdr. html>.

Hewlett-Packard, 1988. Time Domain Reflectometry Theory. Application Note 1304-2; available at <materias.fi.uba.ar/6209/download/HP-AN1304.pdf>.

LeCroy Corp., WaveRunner 6100 Oscilloscope; available at <www.lecroy.com>.

Multicomp. 2N2369 Transistor; available at <www.farnell.com> (distributor).

Sischka, F., 2002. TDR Measurements and Calibration Techniques. Agilent Technologies; available at <eesof.tm.agilent.com/docs/iccap2002/MDLGBOOK/1MEASUREMENTS/71T DR/1TDRCalibration.pdf>.

Williams, J., 1991. High Speed Amplifier Techniques: A Designer's Companion for Wideband Circuitry. Application Note 47. Linear Technology Corp.; available at <www.mit. edu/~6.331/an47fa.pdf>.

Chapter 4

Agilent Technologies, Spectrum Analysis Basics. Application Note AN150; available at <www. metrictest.com/resource_center/pdfs/agl_spec_analyzer_basics.pdf>.

Hameg Instruments. HZ530 Probe set; available at <www.hameg.com>.

Williams, T., 2007. EMC for Product Designers. Newnes.

Chapter 5

Ott, H.W., 1988. Noise Reduction Techniques in Electronic Systems, 2nd ed. John Wiley & Sons Inc.

Chapter 6

Cerna, M., Harvey, A.F., 2000. The Fundamentals of FFT-Based Signal Analysis and Measurement, Application Note 041. National Instruments; available at <zone.ni.com/ devzone/cda/tut/p/id/4278>.

Frigo, M., Johnson, S.G., 2009. FFTW library (developed at MIT); available at <www.fftw.org>.

Chapter 7

Smith III., J.O. FIR Digital Filter Design. DSPrelated.com; available at <www.dsprelated.com/ dspbooks/sasp/FIR_Digital_Filter_Design.html>.

Chapter 8

Altera Corporation, 2007. Understanding CIC Compensation Filters. Application Note 455; available at <www.altera.com/literature/an/an455.pdf>.

Analog Devices, Inc. AD6620 digital receiver.

Donadio, M.P., 2000. CIC Filter Introduction; available at <users.snip.net/~donadio/cic.pdf>.

Lyon, R., 2005. Understanding cascaded integrator-comb filters. In Embedded Systems Programming; available at <www.embedded.com/columns/technicalinsights/160400592?_requestid=511057>.

Microchip Inc. PIC24FJ64GA002 microcontroller.

Chapter 9

Hewlett-Packard. 5372A Frequency and time interval analyzer and 3585 network analyzer; available at <www.hp.com>.

Dove Electronic Components, Inc., Crystal Oscillator Reference; available at <www.doveonline.com/oscillator/moreon-crystal-oscillators.php>.

Fairchild Semiconductor Corp, 1983. HCMOS Crystal Oscillators. Application Note 340. AN005347; available at <www.fairchildsemi.com/an/AN/AN-340.pdf>.

Haque, M., Cox, E., 2004. Use of the CMOS Unbuffered Inverter in Oscillator Circuit. Application Report, SZZA043.Texas Instruments, Inc.; available at <focus.ti.com/lit/an/szza043/szza043.pdf>.

Hewlett-Packard, 1997. Fundamentals of Quartz Oscillators, 200-2-1997; available at <cp.literature.agilent.com/litweb/pdf/5965-7662E.pdf>.

Williamson, T., 1983. Oscillators for Microcontrollers, AP-155. Intel Corp. available at <ftp://download.intel.nl/design/mcs51/applnots/23065901.pdf>.

Voyelle, P., 1985. "Piezoélectricité," In Encyclopedia Universalis, Corpus 14.

Wikipedia. "Crystal oscillator"; available at <en.wikipedia.org/wiki/Crystal_oscillator>.

Chapter 10

Barrett, C., 1999. SWRA029: Fractional/Integer-N PLL Basics. Texas Instruments, Inc.

Brannon, B., 2004. Sampled Systems and the Effects of Clock Phase Noise and Jitter. Application Note AN-756. Analog Devices, Inc.

Analog Devices, Inc., ADIsimPLL V3 Virtual design software; available at <www.analog.com>.

Cypress Semiconductor Corp. CyberClocks online software and CY22393 programmable clock generator; available at <www.cypress.com>.

Labcenter Electronics. Proteus ProSpice simulator; available at <www.labcenter.co.uk>.

Chapter 11

Agilent Technologies, Inc., 2005. Agilent 33220A 20 MHz Function/Arbitrary Waveform Generator Data Sheet.

Analog Devices, Inc., AD9833 Waveform generator, AD9910 direct digital synthesizer, and AD9912 direct digital synthesizer; available at <www.analog.com>.

Analog Devices, Inc., 1999. A Technical Tutorial on Digital Signal Synthesis; available at <www.analog.com/ploadedFiles/Tutorials/450968421DDS_Tutorial_rev12-2-99.pdf>.

Gentile, K., Brandon, D., Harris, T., 2003. DDS Primer. Analog Devices, Inc.; available at <www.ieee.li/pdf/iewgraphs_dds.Pdf>.

Hi-Tech Software. PICC-Lite Compiler; available at <www.htsoft.com>.

Labcenter Electronics. Proteus VSM Mixed-signal simulator; available at <www.labcenter-electronics.com>.

Microchip Technology, Inc. MCP6002 op-amp and PIC16F629A microcontroller; available at <www.microchip.com>.

Chapter 12

Altera Corp. Stratix II GX FPGA. <www.altera.com>.

Agilent Technologies, Inc., 2008. Understanding the effects of limited bandwidth channels on digital data signals. Agilent Measurement Journal Issue 5, May:70–75; available at <www.agilent.com/go/journal>.

INRIA. Scilab. <www.scilab.org>.

Labcenter Electronics. Proteus VSM mixed-signal simulator. <www.labcenter.co.uk>.

Lattice Semiconductor Corp., 2005. Transmission of Emerging Serial Standards over Cable; available at <www.latticesemi.com/lit/docs/generalinfo/Transmission_whitepaper.pdf>.

LeCroy Corp. Waverunner 6050 digital oscilloscope; available at <www.lecroy.com>.

Maxim Integrated Products, Inc., RS-485/RS-422 Transceivers with preemphasis; available at <www.maxim-ic.com>.

Maxim Integrated Products, Inc., 2001. Pre-Emphasis Improves RS-485 Communications. AN643; available at <www.maxim-ic.com/appnotes.cfm/an_pk/643>.

National Semiconductor Corp. EQ50F100n Equalizer, DS64EV100 Equalizer, and DS40MB200 mux/buffer; available at <www.national.com>.

National Semiconductor Corp., 2005. Setting Pre-Emphasis Level for DS40MB200 Dual 4Gb/s Mux/Buffer. AN-1389; available at <www.national.com/an/AN/AN-1389.pdf>.

Stephens, R., 2008. Answering next-gen serial challenges. In Tektronix Knowledge Series, Parts 1–3. Tektronics, Inc.; available at <www.tek.com/Measurement/programs/americas/serialdata_webinar/>.

Tyco Electronics. HSSDC Cable assemblies; available at <www.tycoelectronics.com>.

Wong, W., et al., 2007. Digitally Assisted Adaptive Equalizer in 90 nm with Wide Range Support from 2.5 Gbps to 6.5 Gbps. Altera Corp.; available at <www.altera.com/literature/cp/cp-01026.pdf>.

Zhang, J., Wong, Z., 2002. White Paper on Transmit Pre-Emphasis and Receive Equalization. Mindspeed Technologies/Conexant; available at <www.analogzone.com/io_shoot_mindspeed.pdf>.

Chapter 13

Agilent E4432B digital RF signal generator. <www.agilent.com>.

Agilent E4406A digital transmitter tester. <www.agilent.com>.

Agilent, 2007. Digital Modulation in Communications Systems—An Introduction. Application Note 1298; available at <cp.literature.agilent.com/litweb/pdf/5965-7160E.pdf>.

Bazile, C., Duverdier, A., 2006. First Steps to Use Scilab for Digital Communications. CNES; available at <www.scilab.org/contrib/download.php?fileID=217&attachFileName1=ComNumSc.zip>.

Loy, M., 1999. Understanding and Enhancing Sensitivity in Receivers for Wireless Applications. Technical Brief. SWRA030. Texas Instruments; available at <focus.ti.com.cn/cn/lit/an/swra030/swra030.pdf>.

Langton, C., 2002. Intuitive Guide to Principles of Communications: All about Modulation; available at <www.complextoreal.com/chapters/mod1.pdf>.

McDermott, T., 2006. Wireless Digital Communications: Design and Theory. NEG, Tucson Amateur Packet Radio Corporation; available at <www.tapr.org>.

Chapter 14

Cebik, L.B., Hallidy, D., Jansson, D., Lewallen, R., 2007. The ARRL Antenna Book: The Ultimate Reference for Amateur Radio Antennas, Transmission Lines and Propagation, 21st ed. American Radio Relay League.

Loy, M., Sylla, I., 2005. ISM-Band and Short Range Device Antennas. SWRA046. Texas Instruments, Inc.

Marshall, T., 2002. Numerical Electromagnetics Code; available at <www.nec2.org>.

Miron, D., 2006. *Small Antenna Design*. Newnes.

Reppel, M., Dr. Mühlhaus Consulting & Software, 2005. Analysis of an 868-MHz Loop Antenna., D-58452.

Voors, Arie. 4NEC2 Simulator; available at <home.ict.nl/~arivoors>.

Sonnet Software, Inc., Sonnet Professional and Lite electromagnetic field solvers; available at <www.sonnetsoftware.com>.

Chapter 15

Maxim Integrated Products, Inc. DS1302 Timekeeping chip; available at <www.maxim-ic.com>.

Microchip Technology, Inc. MCP1700 Voltage regulator and PIC16F914 microcontroller; available at <www.microchip.com>.

Microchip Technology, Inc., 2005. PIC16F913/914/916/917/946 Data Sheet: 28/40/44/64-Pin Flash-Based, 8-Bit CMOS Microcontrollers with LCD Driver and nanoWatt Technology. DS41250F. Xxx.

Wikipedia. "World energy resources and consumption"; available at <en.wikipedia.org/wiki/World_energy_resources_and_consumption>.

Chapter 16

Alciom, PowerSpy power analyzer; available at <www.alciom.com>.

Fluke 43B power analyzer; available at <www.fluke.com>.

ST Microelectronics. EVL6566B-40WSTB 40W flyback AC/DC available at <www.st.com/stonline/products/literature/an/14630.pdf>.

ST Microelectronics. L6562A PFC preregulator; available at <www.st.com>.

ST Microelectronics. L6566B flyback controller; available at <www.st.com>.

EU, 2005. Directive 2005/32/EC of the European Parliament and of the Council; available at <www.energy.eu/directives/l_19120050722en00290058.pdf>.

Persson, E., 2006. Understanding PFC. International Rectifier; available at <www2.electronicproducts.com/Understanding_PFC-article-irf-mar2006-html.aspx>.

Chapter 17

LeCroy Corp. WaveRunner 6050A Oscilloscope; available at <www.lecroy.com>.

Wescott, T., 2006. Applied Control Theory for Embedded Systems. Newnes.

Wikipedia. "Control Theory"; available at <en.wikipedia.org/wiki/Control_theory>.

Chapter 18

eZ-Stunt project. 2nd prize in the Driven to Design 2003 contest, sponsored by Circuit Cellar and Zilog Semiconductors; available at <www.circuitcellar.com/zilogdtd2000/second.htm>.

SourceForge. Maxima Computer Algebra System; available at <maxima.sourceforge.net/>.

Index

Printed and bound by CPI Group (UK) Ltd, Croydon, CR0 4YY

03/10/2024

01040345-0002